GOODS OF THE MIND, LLC

Competitive Mathematics Series

for

Gifted Students in Grades 5 and 6

PRACTICE ARITHMETIC
AND
NUMBER THEORY

Cleo Borac, M. Sc.
Silviu Borac, Ph. D.

This edition published in 2013 in the United States of America.

Editing and proofreading: David Borac, B.Mus.
Technical support: Andrei T. Borac, B.A., PBK

Send all inquiries to:

Goods of the Mind, LLC
1138 Grand Teton Dr.
Pacifica
CA, 94044

Competitive Mathematics Series for Gifted Students
Level III (Grades 5 and 6)
Practice Arithmetic and Number Theory

Contents

FOREWORD

The goal of these booklets is to provide a problem solving training ground starting from the earliest years of a student's mathematical development.

In our experience, we have found that teaching how to solve problems should focus not only on finding correct answers but also on finding better solution strategies. While the correct answer to a problem can typically be obtained in several different ways, not all these ways are equally useful for learning how to solve problems.

The most basic strategy is *brute force*. For example, if a problem asks for the number of ways Lila and Dina can sit on a bench, it is easy to write down all the possibilities: Dina, Lila and Lila, Dina. We arrive at this solution by performing all the possible actions allowed by the problem, leaving nothing to the imagination. For this last reason, this approach is called brute force.

Obviously, if we had to figure out the number of ways 30 people could stand in a line, then brute force would not be as practical, as it would take a prohibitively long time to apply.

Using brute force to obtain the correct answer for a simpler problem is not necessarily a useful learning experience for solving a similar problem that is more complex. Moreover, solving problems in a quantitative manner, assuming that the student can transfer simple strategies to similar but more complex problems, is not an efficient way of learning problem solving.

From this simple example, we see that the goal of *practicing* problem solving is different from the goal of problem solving. While the goal of problem solving is to obtain a correct answer, the goal of practicing problem solving is to acquire the ability to develop strategies, generate ideas, and combine approaches that are powerful enough to solve the problem at hand as well as future similar problems.

While brute force is not a useless strategy, it is not a key that opens every door. Nevertheless, there are problems where brute force can be a useful tool. For instance, brute force can be used as a first step in solving a complex problem: a smaller scale example can be approached using brute force to help the problem solver understand the mechanics of the problem and generate ideas for solving the larger case.

All too often, we encounter students who can quickly solve simple problems by applying brute force and who become frustrated when the solving methods they have been employing successfully for years become inefficient once problems increase in complexity. Often, neither the student nor the parent has a clear understanding of why the student has stagnated at a certain level. When the only arrows in the quiver are guess-and-check and brute force, the ability to take down larger game is limited.

Our series of books aims to address this tendency to continue on the beaten path - which usually generates so much praise for the gifted student in the early years of schooling - by offering a challenging set of questions meant to build up an understanding of the problem solving process. Solving problems should never be easy! To be useful, to represent actual training, problem solving should be challenging. There should always be a sense of difficulty, otherwise there is no elation upon finding the solution.

Indeed, practicing problem solving is important and useful only as a means of learning how to develop better strategies. We must constantly learn and invent new strategies while questioning the limitations of the strategies we are using. Obtaining the correct answer is only the natural outcome of having applied a strategy that worked for a particular problem in the time available to solve it. Obtaining the wrong answer is not necessarily a bad outcome; it provides insight into the fallacies of the method used or into the errors of execution that may have occured. As long as students manifest an interest in figuring out strategies, the process of problem solving should be rewarding in itself.

Sitting and thinking in a focused manner is difficult to train, particularly since the modern lifestyle is not conducive to adopting open-ended activities. This is why we would like to encourage parents to pull back from a quantitative approach to mathematical education based on repetition, number of completed pages, and the number of correct answers. Instead, open up the time boundaries that are dedicated to math, adopt math as a game played in the family, initiate a math dialogue, and let the student take his or her time to think up clever solutions.

Figuring out strategies is much more of a game than the mechanical repetition of stepwise problem solving recipes that textbooks so profusely provide, in order to "make math easy." Mathematics is not meant to be easy; it is meant to be interesting.

Solving a problem in different ways is a good way of comparing the merits of each method - another reason for not making the correct answer the primary goal of the activity. Which method is more labor intensive, takes more time or is more prone to execution errors? These are questions that must be part of the problem solving process.

In the end, it is not the quantity of problems solved, the level of theory absorbed, or the number of solutions offered in ready-made form by so many courses and camps, but the willingness to ask questions, understand and explore limitations, and derive new information from scratch, that are the cornerstones of a sound training for problem solvers.

These booklets are not a complete guide to the problem solving universe, but they are meant to help parents and educators work in the direction that, aside from being the most efficient, is the more interesting and rewarding one.

The series is designed for mathematically gifted students. Each book addresses an age range as some students will be ready for this content earlier, others later. If a topic seems too difficult, simply try it again in a couple of months.

FRACTIONS AND DECIMALS

Decimal numbers are:

- *terminating,*
- *non-terminating repeating,*
- *non-terminating mixed,*
- *non-terminating non-repeating.*

Of these, only the first three types can be converted to fractions and are, therefore, *rational* numbers.

The non-terminating non-repeating decimals cannot be converted to fractions and are, therefore, *irrational* numbers.

Numbers such as π, $\sqrt{2}$, or $\sqrt{3}$ are irrational.

The examples that follow show irrational numbers that are based on a pattern of digits that ensures there is no sequence that repeats.

Examples

1. Terminating decimals (rational):

$$\frac{1}{64} = 0.015625$$

$$\frac{3}{16} = 0.1875$$

2. Non-terminating repeating decimals (rational):

$$\frac{7}{11} = 0.63636363\ldots$$

$$\frac{9}{13} = 0.69230769230\ldots$$

3. Non-terminating mixed decimals (rational):

$$\frac{5}{6} = 0.83333\ldots$$

$$\frac{9}{14} = 0.64285714285\ldots$$

4. Non-terminating non-repeating decimals (irrational):

$$0.1010010001000010\ldots$$

$$1.2122122212222122222\ldots$$

Converting Terminating Decimals into Fractions

Terminating decimal numbers can be converted into fractions by multipliying them by a unitary fraction with a sufficiently large power of 10 at both the numerator and denominator, as follows:

$$0.89 \; = \; 0.89 \times \frac{100}{100} = \frac{89}{100}$$

$$1.924 \; = \; 1.924 \times \frac{1000}{1000} = \frac{1924}{1000}$$

$$5.6 \; = \; 5.6 \times \frac{10}{10} = \frac{56}{10}$$

Note: this process does not guarantee that the result is an *irreducible* fraction.

Identifying a Fraction as a Terminating Decimal

First, the fraction must be simplified until an equivalent irreducible fraction is obtained. The irreducible fraction will be a terminating decimal if its denominator has only factors of 2 and/or 5 in its prime factorization, like in the following example:

$$\frac{60}{75} = \frac{2^2 \times \cancel{3} \times \cancel{5}}{\cancel{3} \times \cancel{5} \times 5} = \frac{4}{5} = \frac{8}{10} = 0.8$$

In the example above, one can notice how:

- the denominator has factors of 2, 3 and 5 before simplification;
- the denominator has only factors of 5 after simplification;
- with only factors of 2 and/or 5 at the denominator, we can always turn the denominator into a power of 10.

Converting fractions to terminating decimals can be done either by using the process shown above, or by performing long division.

Converting Non-Terminating Repeating Decimals into Fractions and Conversely

If the irreducible form of the fraction has prime factors other than 2 and/or 5 in the factorization of its denominator, then it is going to be a non-terminating repeating decimal.

We can find the repeating decimals by performing long division.

A non-terminating repeating decimal consists of a sequence of digits, called a *repetend*, that repeats indefinitely:

$$1.\underbrace{2396714508}_{\text{repetend}} 23967145082396714508\ldots$$

The number of digits in the repetend is called a *period*. The decimal in the example has a period of 10.

To convert a non-terminating repeating decimal into a fraction, we use the idea that, by taking away one occurence of the repetend out of an infinity of occurences, there is still an infinite number of them left:

$$N = 0.934934\ldots$$

$$1000 \times N = 934.934934\ldots$$

Now subtract:

$$1000 \times N - N = 934.934934\ldots - 0.934934\ldots$$

$$999 \times N = 934$$

$$N = \frac{934}{999}$$

12

Converting Mixed Non-Terminating Decimals into Fractions and Conversely

If a fraction has, in its irreducible form, factors of 2 or 5 as well as other prime factors in the factorization of its denominator, then it has a decimal form that is a mixed non-terminating decimal. For example:

$$\frac{231}{65} = \frac{3 \times 7 \times 11}{5 \times 13} = 3.5 \underbrace{538451}_{\text{repetend}} 538\ldots$$

The digit 5 that is bolded is called a *fixed part*.

To convert a mixed decimal into a fraction, we subtract out the repeating decimals:

$$N = 0.36752752\ldots$$

$$100 \times N = 36.752752\ldots$$

$$100000 \times N = 36752.752752\ldots$$

Now subtract:

$$100000 \times N - 100 \times N = 36752.752752\ldots - 36.752752\ldots$$

$$99900 \times N = 36716$$

$$N = \frac{36716}{99900}$$

Continued fractions can be used to represent both rational and irrational numbers.

- All the numerators of a continued fraction are equal to 1.
- The continued fraction of a rational number is finite.
- The continued fraction of an irrational number is infinite.
- The continued fraction of an irrational square root is infinite but periodic.

A continued fraction is derived by applying the division algorithm repeatedly:

$$\frac{15}{4} = 3 + \frac{3}{4}$$

$$= 3 + \frac{1}{\frac{4}{3}}$$

$$= \mathbf{3} + \cfrac{1}{\mathbf{1} + \cfrac{1}{3}}$$

The bolded numbers are called *partial quotients*.

Experiment

Convert the following into continued fractions:

$$\frac{20}{7}, \frac{37}{9}, \frac{91}{11}$$

Infinite continued fractions will be studied in level IV.

The *last digit* of any product of integers is given by the last digit of the product of their last digits:

$$\mathbf{68} \times \mathbf{57} = 38\mathbf{96}$$

This can be proven by using the expanded form of integers and the distributive property of multiplication over addition:

$$(6 \times 10 + 8) \times (5 \times 10 + 7) = 6 \times 5 \times 100 + (6 \times 7 + 5 \times 8) \times 10 + 8 \times 7$$

The last digit of an integer power forms a sequence of repeating blocks of terms: 2^0 ends in 1, 2^1 ends in 2, 2^2 ends in 4, 2^3 ends in 8, 2^4 ends in 6 ....

The sequence of the last digit of the integer powers of 2 is, therefore:

$$1, \ 2, \ 4, \ 8, \ 6, \ 2, \ 4, \ 8, \ 6, \ 2, \cdots$$

Similarly, the sequences of last digits for other integer powers are:

$$3^k \ : \ 1, \ 3, \ 9, \ 7, \ 1, \ 3, \ 9, \ \cdots$$
$$4^k \ : \ 1, \ 4, \ 6, \ 4, \ 6, \ 4, \ 6, \ \cdots$$
$$5^k \ : \ 1, \ 5, \ 5, \ 5, \ 5, \ 5, \ 5, \ \cdots$$
$$6^k \ : \ 1, \ 6, \ 6, \ 6, \ 6, \ 6, \ 6, \ \cdots$$
$$7^k \ : \ 1, \ 7, \ 9, \ 3, \ 1, \ 7, \ 9, \ \cdots$$
$$8^k \ : \ 1, \ 8, \ 4, \ 2, \ 6, \ 8, \ 4, \ \cdots$$
$$9^k \ : \ 1, \ 9, \ 1, \ 9, \ 1, \ 9, \ 1, \ \cdots$$

PRACTICE ONE

Do not use a calculator for any of the problems!

Exercise 1

At a glance (wihout dividing), for each fraction:

- circle T if the fraction is equivalent to a terminating decimal
- circle NTR if the fraction is equivalent to a non-terminating repeating decimal
- circle NTM if the fraction is equivalent to a non-terminating mixed decimal

$\frac{11}{8}$	T NTR NTM		$\frac{123}{144}$	T NTR NTM		$\frac{18}{144}$	T NTR NTM
$\frac{3}{256}$	T NTR NTM		$\frac{121}{275}$	T NTR NTM		$\frac{11}{36}$	T NTR NTM
$\frac{7}{49}$	T NTR NTM		$\frac{693}{132}$	T NTR NTM		$\frac{5}{14}$	T NTR NTM

Exercise 2

Sort in decreasing order:

$$a = 6.538, \ b = 6.5\overline{38}, \ c = 6.\overline{538}, \ d = 6.53\overline{7}$$

(A) $a > d > b > c$

(B) $b > c > d > a$

(C) $c > b > a > d$

(D) $b > d > c > a$

Exercise 3

If k is a positive integer, for which value of k is the fraction below equivalent to the largest possible non-terminating repeating decimal:

$$\frac{14}{7 \times \heartsuit}$$

(A) 1

(B) 2

(C) 3

(D) 5

Exercise 4

k, q, p, and m are positive integer values that make the fraction below equivalent to the largest possible proper, non-integer, non-terminating repeating decimal:

$$\frac{2^4 \times 5^3 \times 7^2 \times 3^5}{2^k \times 5^q \times 7^p \times 3^m}$$

Find $k + q + p + m$.

Exercise 5

Convert the following decimals into irreducible fractions:

$0.\overline{14}$	$1.\overline{5}$	$0.2\overline{34}$	$0.7\overline{896}$	$0.34\overline{5}$	$0.451\overline{3}$	$0.91\overline{28}$

Exercise 6

Find the last non-zero decimal of the division:

$$\frac{17}{5^{2013}}$$

Exercise 7

For which positive integers k is the value of the fraction:

$$\frac{2013}{2^k}$$

a decimal that ends in 5?

(A) only $k = 1$

(B) any k that is a multiple of 4 plus 1

(C) any k that is a multiple of 4 plus 1 or plus 3

(D) any k

Exercise 8

When converting the following fraction to a decimal, which digit is in the 64th decimal place?

$$\frac{1}{7}$$

Exercise 9

Calculate the value of the expression:

$$\cfrac{1}{3 - \cfrac{1}{1 - \cfrac{1}{1 + 0.\overline{6}}}}$$

Exercise 10

Find the digit k if it is true that:

$$5.\overline{k} = 6$$

Exercise 11

What is the value of the integer m?

$$\frac{1}{0.\overline{63}} = \frac{m + 4}{m}$$

Exercise 12

There are integer values m, a, b, c, and d so that:

$$\frac{67}{155} = m + \cfrac{1}{a + \cfrac{1}{b + \cfrac{1}{c + \cfrac{1}{d}}}}$$

What is the value of $m + a + b + c + d$?

Exercise 13

If k is a 2-digit number, what percentage of the fractions:

$$\frac{1}{k}$$

are equivalent to terminating decimals?

Exercise 14

A rectangle has side lengths 19 and 47. Lila cuts it to obtain the largest possible square and another rectangle. She then continues to cut the remaining rectangle in the same way until there is no rectangle left. How many squares does she have at the end?

Exercise 15

The number $0.2981736451\ldots$:

(**A**) is rational

(**B**) is irrational

(**C**) could be either rational or irrational

(**D**) is neither rational nor irrational, since it is decimal

INTEGER DIVISION

The Integer Division Theorem

For any two integers D and d, with $d > 0$, there is a *unique* pair of integers Q and r, with $r < d$, such that:

$$D = d \times Q + r$$

D is the *dividend*, d is the *divisor*, Q is the *quotient*, and r is the *remainder*.

Important Facts:

- The divisor cannot be zero ($d \neq 0$).

- The remainder must be smaller than the divisor. Otherwise, the division can continue.

- There is a finite set of remainders for a given divisor: $\{0, 1, 2, \ldots, (d - 1)\}$.

- The number of possible remainders equals the divisor (there are d numbers from 0 to $(d - 1)$).

- If D, the dividend, is zero, the quotient and the remainder are zero as well. In this case, the divisor can have any non-zero value.

Example:

$$15 = 3 \times 4 + 3$$

is an integer division by 4. It cannot be construed as a division by 3 since the remainder is 3.

A *prime* number is a positive integer that has only two divisors: 1 and itself.

A *composite* number is a positive integer that has more than two divisors.

The number 1 is neither prime, nor composite - it is an *improper prime*.

The Fundamental Theorem of Arithmetic

Any positive integer N greater than 1 is either prime or composite. If composite, it has a *unique* factorization into primes:

$$N = p_1^{a_1} \times p_2^{a_2} \times \cdots p_k^{a_k}$$

where $p_1 < p_2 < \cdots < p_k$ are distinct primes ordered increasingly, and $a_1, a_2, \ldots, a_k$ are positive exponents.

From now on, we will write prime factorizations in this way, using the largest possible integer exponent for each of the primes involved.

Example:

$$12,600 = 2^3 \times 3^2 \times 5^2 \times 7$$

Testing for Primality

Problems often require finding out whether a number is prime. To answer this question, one *must* perform all the divisions by prime numbers up to the prime that is closest to the square root of the given number.

Example

Is 6731 prime?

First, estimate the square root of 6731. Since $80 \times 80 = 6400$, the square root is close to 80. Further inquiry shows that:

$$82 \times 82 < 6731 < 83 \times 83$$

Therefore, we have to consider dividing by all the primes smaller than 83.

Next, apply various simple divisibility tests to rule out some of the divisions:

- the number is odd, therefore 2 is not a factor
- the digit sum is $6 + 7 + 3 + 1 = 17$, therefore the number is not divisible by 3
- the last digit is 1, therefore it is not divisible by 5
- the alternating digit sum is $1 - 3 + 7 - 6 = -1$, therefore the number is not divisible by 11

From here, we have to divide by all the primes in increasing order, up to 79. It's a tough job, but it must be done. We find that:

$$53 \times 127 = 6731$$

PRACTICE TWO

Do not use a calculator for any of the problems!

Exercise 1

Dividing a 2-digit positive integer by 17 produces a remainder of 9. What is the difference between the smallest such integer and the largest such integer?

Exercise 2

What is the sum of the integer numbers that produce a quotient of 19 when divided by 11?

Exercise 3

Which of the following choices is the irreducible form of the fraction:

$$\frac{12155}{17017}$$

(A) $\dfrac{5}{7}$

(B) $\dfrac{1105}{1547}$

(C) $\dfrac{85}{119}$

(D) $\dfrac{12155}{17017}$

Exercise 4

If the product of three consecutive numbers has a factor of 3^6, what is the smallest possible value for the smallest of the three numbers?

(A) 3^2

(B) 3^3

(C) $3^2 - 2$

(D) $3^6 - 2$

Exercise 5

If the product of three consecutive numbers has a factor of 2^9, what is the smallest possible value for the smallest of the three numbers?

Exercise 6

Factor into primes:

$$847 =$$

$$8189 =$$

$$6754 =$$

$$4345 =$$

$$2132 =$$

$$4947 =$$

Exercise 7

The positive difference (*positive difference* means that we subtract the smaller number from the larger number) of the squares of two consecutive numbers is:

(A) always odd

(B) always even

(C) sometimes odd

Exercise 8

Find three prime numbers a, b, and c such that:

$$a + b + c = 516$$

Exercise 9

Find the prime numbers k and p such that:

$$2^k + p = 151$$

Exercise 10

Two consecutive numbers, p and k, are prime. Find:

$$k^p + p^k$$

Exercise 11

Write 3^5 as a sum of three consecutive numbers. What is the largest prime divisor among the prime divisors of the three numbers?

(A) 1

(B) 2

(C) 3

(D) 41

Exercise 12

Write 2^7 as a sum of consecutive odd numbers.

Exercise 13

The remainder from dividing N by 34 is 31. What is the remainder from dividing N by 17?

Exercise 14

Write the number 33306 as a product of two consecutive numbers.

Divisors and Divisibility

A positive integer N has the following kinds of *divisors*:

Prime divisors: The set of primes that divide N. If $N = 75$, the prime divisors are 3 and 5.

Positive divisors: The set of positive integers that divide N. If $N = 75$, the positive divisors are: 1, 3, 5, 15, 25, and 75.

Proper divisors: The set of proper divisors is the set of positive divisors that are different from N. If $N = 75$, the proper divisors are: 1, 3, 5, 15 and 25.

Integer divisors: The set of integer divisors includes both positive and negative divisors. If $N = 75$, the set of integer divisors is: $-75 - 25$, -15, -5, -3, -1, 1, 3, 5, 15, 25, and 75.

Experiment

Write out all the sets of divisors of the number 52.

Solution:

The prime factorization of 52 is $2^2 \times 13$. From here we find:

Prime divisors: $\{2, 13\}$

Positive divisors: $\{1, 2, 4, 13, 26, 52\}$

Proper divisors: $\{1, 2, 4, 13, 26\}$

Integer divisors: $\{\pm 1, \pm 2, \pm 4, \pm 13, \pm 26, \pm 52\}$

Divisibility by powers of 2 or 5: 2, 5, 4, 25, <u>8</u>, etc.

A power of 2 will divide an integer if, by separating as many digits as the power from the rightmost part of the number, we obtain a number divisible by the given power of 2.

For 2:
$$123456 = 123450 + 6 = 12345 \cdot 2 \cdot 5 + 6$$

The number 123450 is divisible by 2 because it is a multiple of 10. If the last digit is divisible by 2, the larger number is also divisible by 2.

For 4:
$$123456 = 123400 + 56 = 1234 \cdot 2^2 \cdot 5^2 + 56$$

The number 123400 is divisible by 4 because it is a multiple of 100. If the last two digits form a number divisible by 4, the larger number is also divisible by 4.

For 8:
$$123456 = 123000 + 456 = 123 \cdot 2^3 \cdot 5^3 + 456$$

The number 123000 is divisible by 8 because it is a multiple of 1000. If the last three digits form a number divisible by 8, the larger number is also divisible by 8.

Since 5 is also a divisor of 10, exactly the same reasoning applies to divisibility by powers of 5 (5, 25, 125.)

Divisibility by 9 or by 3

Rules of divisibility by 9 and by 3 can be derived by casting out multiples of 9. For example:

$$
\begin{aligned}
5234 &= 4 + 3 \cdot 10 + 2 \cdot 100 + 5 \cdot 1000 \\
&= 4 + 3 \cdot (9 + 1) + 2 \cdot (99 + 1) + 5 \cdot (999 + 1) \\
&= 9 \cdot (3 + 2 \times 11 + 5 \times 111) + 4 + 3 + 2 + 5
\end{aligned}
$$

We see that any number is the sum of a multiple of 9 and its digit sum. If the digit sum is a multiple of 9, the number is a multiple of 9 as well.

If the digit sum is not a multiple of 9 but is a multiple of 3, then the whole number is a multiple of 3 but not of 9.

Casting out multiples of 9 is particularly easy because 9 differs by 1 from 10, the base of numeration.

Note: Any rule of divisibility is valid only for a specific base of numeration.

Divisibility by 11

A rule of divisibility by 11 can be derived by casting out multiples of 11. For example:

$$
\begin{aligned}
5234 &= 4 + 3 \cdot 10 + 2 \cdot 100 + 5 \cdot 1000 \\
&= 4 + 3 \cdot (11 - 1) + 2 \cdot (99 + 1) + 5 \cdot (1001 - 1) \\
&= 4 + 3 \cdot 11 - 3 + 2 \cdot 9 \cdot 11 + 2 + 5 \cdot 91 \cdot 11 - 5 \\
&= 11 \cdot (3 + 2 \times 9 + 5 \times 91) + 4 - 3 + 2 - 5
\end{aligned}
$$

We see that any number is the sum of a multiple of 11 and its alternating digit sum. If the alternating digit sum is a multiple of 11, the number is a multiple of 11 as well.

Casting out multiples of 11 is particularly easy because 11 differs by 1 from 10, the base of numeration.

Note: Any rule of divisibility is valid only for a specific base of numeration.

Divisibility by 7 and 13

Mini-rules of divisibility by 7 and 13 can be derived by using the fact that $1001 = 7 \cdot 11 \cdot 13$. For example,

$$
\begin{aligned}
5226 &= 226 + 5 \cdot 1000 \\
&= 226 + 5 \cdot (1001 - 1) \\
&= 225 + 5 \cdot 1001 - 5 \\
&= 5 \cdot 1001 + 221
\end{aligned}
$$

If 221 is divisible by 7 (or by 13), then the number 5226 is also divisible by 7 (or by 13).

PRACTICE THREE

Do not use a calculator for any of the problems!

Exercise 1

List the proper divisors of 531.

Exercise 2

What is the largest odd divisor of 504?

Exercise 3

How many positive integers smaller than 1000 have exactly 4 proper divisors?

(A) 1

(B) 2

(C) 3

(D) more than 3

Exercise 4

What is the remainder if we divide 540 by the sum of its positive divisors?

Exercise 5

The number 2013 can be written as a sum of k consecutive integers. How many of the following numbers are valid values for k?

$$11, \ 29, \ 33, \ 57, \ 61, \ 73, \ 99, \ 183, \ 671$$

(A) 3

(B) 5

(C) 8

(D) 9

Exercise 6

If n is a positive integer greater than 1, what are the known numeric divisors of the expression:

$$(n-1)n(n+1)$$

Exercise 7

If n is a positive integer and the largest power of 2 that divides the expression $E = (n-1)n(n+1)$ is 2^{12}, how many of the following statements could be true?

  i. Each of $n-1$, n, and $n+1$ must be divisible by 8.

  ii. Each of $n-1$, n, and $n+1$ must be divisible by 16.

  iii. n is divisible by 2^{12}.

  iv. $n-1$ is divisible by 2^{12}.

  v. Either $n-1$ or $n+1$ must be divisible by 2^{11}, while the other must be divisible by 2.

Exercise 8

When dividing k positive consecutive numbers by 5, the sum of the remainders is 31. What is the product of the remainders?

(A) 0

(B) 1

(C) 2

(D) 24

Exercise 9

If a and b are randomly chosen non-zero digits, which of the following numbers is a divisor of $baab$?

(A) 5

(B) 7

(C) 11

(D) 13

Exercise 10

What is the sum $a + b + c$ if a, b, and c are different digits that make the number $7a5bc$ divisible by 2, by 5, and by 11?

(A) 0

(B) 7

(C) 12

(D) 23

Exercise 11

What is the remainder of the integer division $a \div b$ if:

$$a = 1 + 2 + 3 + \cdots + 90$$

and

$$b = 2 + 4 + 6 + \cdots + 12$$

Exercise 12

The 2-digit integer ab has different digits a and b such that a is twice b. What is the sum of all such 2-digit numbers?

Exercise 13

Which of the following are divisors of any repdigit with 5 digits? (A *repdigit* is an integer written using one digit repeatedly.) Check all that apply.

(A) 5

(B) 11

(C) 41

(D) 101

Exercise 14

If a, b, and c are digits, how many of the following numbers must be divisors of $abcabcabc$?

$$abc, \ 111, \ 1001001, \ 3, \ 333667, \ a00a00a00, \ bc, \ 9, \ 1001$$

(A) 1

(B) 3

(C) 4

(D) 5

Exercise 15

What is the largest 5-digit palindrome divisible

    i. by 8

   ii. by 5

  iii. by 11

Exercise 16

An integer has 111 digits of 4 and 111 digits of 7. Is there an arrangement of the digits that makes this number a perfect square?

Exercise 17

If n is a non-negative integer, the number of zeros at the end of $n!$ forms the sequence:

$$0, \ 0, \ 0, \ 0, \ 0, \ 1, \ 1, \ 1, \ 1, \ 1, \ 2, \ 2, \ 2, \ldots$$

The 26th term of this sequence is:

(A) 3

(B) 4

(C) 5

(D) 6

Exercise 18

The 3-digit number A produces a remainder of 3 when divided by 5, and a remainder of 6 when divided by 8. What is the smallest possible value of A?

Exercise 19

To reduce the amount of work needed for testing whether a three digit number is divisible by 8, we cast out 8s as follows. Let us assume a 3-digit number abc:

$$
\begin{aligned}
abc &= ab \cdot 10 + c \\
&= ab \cdot (2 + 8) + c \\
&= 2 \cdot ab + c + 8 \cdot ab
\end{aligned}
$$

By casting out the 8s we see that the remainder from dividing abc by 8 is the same as the remainder from dividing $2 \cdot ab + c$ by 8. If we are not interested in the value of the remainder, but just in whether it is zero or not, we say further that abc is divisible by 8 if $ab + \frac{c}{2}$ is divisible by 4.

Test this method on each of the following numbers and circle the ones that are divisible by 8:

$$224, \ 136, \ 482, \ 864, \ 942, \ 748$$

Exercise 20

Find the digits b and c which make the number $4b5c$ a multiple of 72.

Bases of Numeration

A *base of numeration* is the number of digits available to symbolize numbers in a positional system of numeration.

The *positional system* makes use of place values. The Roman system of numeration is not positional and has, therefore, no base of numeration.

Bases used nowadays include:

Base 10 *(decimal numbers)* is used in daily life.

Base 60 *(sexagesimal numbers)* is used to express the time of day and the measures of angles.

Base 2 *(binary numbers)* is used in digital circuitry.

Base 8 *(octal numbers)* is used by computer engineers.

Base 16 *(hexadecimal numbers)* is used by computer engineers.

Notation: Bases are usually symbolized as a number written at the subscript level. Examples:

- 15_8 is one five in base eight;

- 211_3 is two, one, one in base three

- $AB1_{14}$ is A, B, one in base 14

Note that we cannot use words like "hundreds" or "tens" when we read numbers in bases other than 10.

Converting a Number from Some Base to Base 10

Example 1: Assume the digits of base 6 are: 0, 1, 2, 3, 4, and 5. Convert the number 123_6 to decimal notation by writing it in expanded form:

$$123_6 = 3 + 2 \times 6 + 1 \times 6^2 = 3 + 12 + 36 = 51_{10}$$

Example 2: Assume the digits of base 3 are 0, 1, and 2. Convert the number 2221_3 to decimal notation by writing it in expanded form:

$$
\begin{aligned}
2221_3 &= 1 + 2 \times 3 + 2 \times 3^2 + 2 \times 3^3 \\
2221_3 &= 1 + 6 + 18 + 54 \\
2221_3 &= 79_{10}
\end{aligned}
$$

Converting a Number from Base 10 to Another Base

1. Divide the number by the base.

2. Divide the quotient from the previous division by the base.

3. Repeat Step 2 until the quotient becomes zero.

4. Collect all the remainders in reverse order (from last to first) - this is the result.

Example: Convert 49_{10} to base 4:

$$
\begin{aligned}
49 &= 4 \times 12 + 1 \\
12 &= 4 \times 3 + 0 \\
3 &= 4 \times 0 + 3 \\
49_{10} &= 301_4
\end{aligned}
$$

Writing a Number as a Sum of Different Powers of 2

Any positive integer can be written as a sum of different powers of 2 because it has a base 2 representation. The digits in this base are 0 and 1 - their behavior is to turn "on" or "off" some power of 2. This does not work in other bases.

Example 1 Write 25 as a sum of powers of 2.

Notice that it is sufficient to convert the number to binary. Write the integer divisions and collect the remainders in reverse order:

$$
\begin{aligned}
25 &= 12 \times 2 + 1 \\
12 &= 6 \times 2 + 0 \\
6 &= 3 \times 2 + 0 \\
3 &= 1 \times 2 + 1 \\
1 &= 0 \times 2 + 1
\end{aligned}
$$

Therefore:
$$25 = 11001_2 = 2^4 + 2^3 + 2^0$$

Example 2 An old word problem says: "What is the smallest number of weights we need in order to measure any integer mass between 1g and 100g?"

It is sufficient to convert 100 to binary: 1100100_2 to find that a minimum of 7 weights are needed: 64g, 32g, 16g, 8g, 4g, 2g, and 1g. Any weight between 1 g and 100g can be expressed as a base 2 number - just use the weights selected by the digits of 1.

Writing a Number as an Algebraic Sum of Different Powers of 3

Any positive integer can be written as an algebraic sum (sum where terms can be positive and negative integers) of powers of 3, because any number has a base 3 representation.

Strategy 1: using a base 3 representation

In base 3, there are only 3 digits: 0, 1, and 2. The digits 0 and 1 "turn on" and "off" powers of 3. However, the digit 2 is not so convenient. We can write the digit 2 as $3 - 1$. Therefore, the action of a digit of 2 is as follows:

$$
\begin{aligned}
222_3 &= 2 \times 3^2 + 2 \times 3 + 2 \times 3^0 \\
&= (3-1) \times 3^2 + (3-1) \times 3 + 3 - 1 \\
&= 3^3 - 3^2 + 3^2 - 3 + 3 - 1 \\
&= 3^3 - 1
\end{aligned}
$$

Ad-hoc Strategy: The same result can be obtained, for this particular example, by noticing that 222_3 is like 999 in base 10. That is: $1000_3 - 1 = 222_3$, which is equivalent to:

$$
222_3 = 1000_3 - 1 = 3^3 - 1
$$

Example Write 45 as an algebraic sum of powers of 3. Start by converting the number to base 3:

$$45 = 15 \times 3 + 0$$
$$15 = 5 \times 3 + 0$$
$$5 = 1 \times 3 + 2$$
$$1 = 0 \times 3 + 1$$

Therefore, $45 = 1200_3$. Write in expanded form:

$$
\begin{aligned}
1200_3 &= 3^3 + 2 \times 3^2 \\
&= 3^3 + (3-1) \times 3^2 \\
&= 3^3 + 3^3 - 3^2 \\
&= 2 \times 3^3 - 3^2 \\
&= (3-1) \times 3^3 - 3^2 \\
&= 3^4 - 3^3 - 3^2
\end{aligned}
$$

Strategy 2: using base 3 with negative digits

The same result can be obtained by introducing the concept of a *negative remainder* and using the same algorithm as usual for converting to base 3. To work with a negative remainder, just increase the quotient by 1 in the usual division algorithm.

Example Convert 26 to base 3 using the digits -1, 0, and 1:

$$
\begin{aligned}
26 &= 9 \times 3 - 1 \\
9 &= 3 \times 3 + 0 \\
3 &= 3 \times 1 + 0 \\
1 &= 3 \times 0 + 1
\end{aligned}
$$

The representation of 26 in base 3 using the digits -1, 0, and 1 is:

$$26 = 3^3 - 1$$

Example Convert 103 to an algebraic sum of powers of 3. First, write the number in base 3 using the digits -1, 0, and 1:

$$
\begin{aligned}
103 &= 34 \times 3 + 1 \\
34 &= 11 \times 3 + 1 \\
11 &= 4 \times 3 - 1 \\
4 &= 1 \times 3 + 1 \\
1 &= 0 \times 3 + 1 \\
103 &= 11(-1)11_3
\end{aligned}
$$

Therefore:
$$103 = 3^4 + 3^3 - 3^2 + 3 + 3^0$$

PRACTICE FOUR

Do not use a calculator for any of the problems!

Exercise 1

Convert the following decimal numbers to base 8 (octal):

$$56, \ 89, \ 105, \ 156, \ 203, \ 304, \ 392$$

Exercise 2

Convert the following decimal numbers to base 16 (hexadecimal):

$$41, \ 92, \ 107, \ 203, \ 410, \ 548, \ 619$$

Exercise 3

Convert the following decimal numbers to base 2 (binary):

$$14, \ 21, \ 25, \ 30, \ 65, \ 78, \ 91$$

Exercise 4

Convert the following numbers to base 10 (decimal):

$$54_8, \ 88_9, \ A1_{16}, \ BBA_{13}, \ 101_3, \ 214_6, \ 337_8$$

Exercise 5

For which base b is the following equality true?

$$3 \times 4 = 15$$

Exercise 6

Write the following numbers as sums of powers of 2:

$$45, \ 52, \ 63, \ 71, \ 89, \ 115, \ 296, \ 3012$$

Exercise 7

Write the following numbers as algebraic sums of powers of 3 (sums where terms can be both negative and positive):

$$45, \ 52, \ 63, \ 71, \ 89, \ 115, \ 296$$

Exercise 8

$0.\overline{3}$ is a number in base 7. Its representation in base 10 is:

(A) 0.5

(B) $0.\overline{6}$

(C) $0.\overline{428571}$

(D) $0.\overline{3}$

Exercise 9

What is the result of the operation $100000 - 1$:

(a) in base 3?

(b) in base 5?

(c) in base 11?

Exercise 10

The number n^2 is written as 80 in base k. What is $8k$ in base n?

(A) 8

(B) 10

(C) 80

(D) 100

Exercise 11

The elements of set S are the digit sums of all the 3-digit numbers in base 12. How many elements does the set S have?

Exercise 12

Find the base x:

$$17_x + 24_x = 50_{x-1}$$

Exercise 13

Which of the following binary numbers is not divisible by 4?

(A) 1010101010100

(B) 1000100011100

(C) 1110011001000

(D) 1010100001010

Exercise 14

Let us extend the concept of base to numbers with decimals. An example of the expanded form of a decimal number in base 10 is:

$$231.14 = 2 \times 10^2 + 3 \times 10 + 1 + 1 \times \frac{1}{10} + 4 \times \frac{4}{10^2}$$

Convert the number 2.6 to base 5.

LCM and GCF

The *least common multiple* (LCM) of a set of positive integers is the smallest integer that is divisible by each of the numbers.

To compute the LCM:

1. Factor all the numbers into primes.

2. Multiply together all the distinct primes that can be seen in all factorizations.

3. Raise each prime in the product to the largest power it is at in the various factorizations. This is the LCM.

Example 1: Compute the LCM of the numbers $36, 24$, and 48.

$$
\begin{aligned}
36 &= 2^2 \times 3^2 \\
24 &= 2^3 \times 3 \\
48 &= 2^6 \times 3 \\
LCM(36, 24, 48) &= 2^6 \times 3^2 = 576
\end{aligned}
$$

Coprime numbers are numbers that do not have any factors in common, except 1.

The LCM of two coprime numbers is equal to the product of the numbers.

The *greatest common divisor/factor* (GCD or GCF) of a set of positive integers is the largest integer that is a factor of each of the numbers. To compute the GCD:

1. Factor all the numbers into primes.

2. Multiply together only the primes that appear in all factorizations.

3. Raise each prime in the product to the smallest power it is at in the various factorizations. This is the GCD.

Example 1: Compute the GCD of the numbers $36, 24,$ and 48.

$$
\begin{aligned}
36 &= 2^2 \times 3^2 \\
24 &= 2^3 \times 3 \\
48 &= 2^6 \times 3 \\
GCD(36, 24, 48) &= 2^2 \times 3 = 12
\end{aligned}
$$

The product of the LCM and GCF of two positive integers is equal to the product of the integers:

$$
\text{lcm}(a, \ b) \times \text{gcf}(a, \ b) = a \times b
$$

Since the GCF contains factors that are common to both numbers, it is a divisor of the LCM. By multiplying the LCM and the GCF the common factors will appear twice, while the factors that are not common will appear once - this is equivalent to multiplying the two numbers together.

Practice Five

Exercise 1

Find the greatest common factor (greatest common divisor) of each of the following sets of numbers or expressions:

1. $\gcf(1008,\ 234,\ 1116)$
2. $\gcf(2^5 \times 37^2 \times 5,\ 2^3 \times 5^2 \times 7)$
3. $\gcf(a^2b^3c^4,\ a^3b^2c^3,\ a^4b^3c^2)$
4. $\gcf(2^6 \times s^3,\ 2^8 \times s^2)$

Exercise 2

Find the least common multiple of each of the following sets of numbers or expressions:

1. $\lcm(155,\ 713)$
2. $\lcm(1000, 1400)$
3. $\lcm(a^2b^3c^4,\ a^3b^2c^3,\ a^4b^3c^2)$
4. $\lcm(385,\ 98,\ 55)$

Exercise 3

When divided by 4, by 5, or by 6, the number N produces the remainder 3. What is the smallest possible value of N?

Exercise 4

A positive integer M produces a remainder of 3 when divided by 7, a remainder of 4 when divided by 8, and a remainder of 5 when divided by 9. How many such numbers are there from 1 to 1000, inclusive?

(A) 0

(B) 1

(C) 2

(D) 3

Exercise 5

The product of the lcm and the gcf of two positive integers larger than 1 is 391. What is the positive difference of the integers?

Exercise 6

The positive integers N and M have $\operatorname{lcm}(N,\ M) = 5670$ and $\gcd(N,\ M) = 18$. What is the smallest possible value of $N + M$?

Exercise 7

Find the smallest sum $m + n + p$ such that:

$$
\begin{aligned}
\gcd(m,\ n) &= 12 \\
\gcd(m,\ p) &= 30 \\
\gcd(p,\ n) &= 18
\end{aligned}
$$

Exercise 8

Find the smallest number that, when divided by 14 produces a remainder of 9, when divided by 12 produces a remainder of 7, and when divided by 21 produces a remainder of 16.

Exercise 9

If m and n are two positive integers that satisfy $m^2 + n^2 = 8281$ and $\gcd(m,\ n) = 7$, how many different pairs $(m,\ n)$ are there?

Exercise 10

If $\text{lcm}(m, n) + \text{gcf}(m, n) = 113$, where m and n are two positive integers larger than 1, find $m + n$.

Exercise 11

When dividing a positive integer N by 12, we get a remainder of 5 and when dividing the same integer by 13, we get a remainder of 9. What is the remainder when dividing N by 12×13?

Exercise 12

How many positive integers smaller than 5000 produce a remainder of 7 when divided by 18, 21, and 30?

(A) 0

(B) 2

(C) 4

(D) 7

Exercise 13

If a and b are positive integers with $\text{lcm}(a, b) = 840$ and $\text{gcf}(a, b) = 28$, find $a \times b$.

Exercise 14

If $\text{lcm}(a, b) = 2^2 \cdot 3^2$ and $\text{gcf}(a, b) = 3$, which of the following must be true:

(A) a and b are coprime.

(B) a and b are both even.

(C) a and b are both odd.

(D) a and b have different parity.

PERFECT SQUARES AND CUBES

A *perfect square* is a positive integer that can be written as the product of an integer by itself.

A *perfect cube* is an integer that can be written as the product of an integer and its square.

A list of perfect squares and cubes of the numbers from 1 to 9:

1	2	3	4	5	6	7	8	9
1	4	9	16	25	36	49	64	81
1	8	27	64	125	216	243	512	729

Facts about perfect squares:

1. Perfect squares never end in 2, 3, 7, or 8.

2. Perfect squares have only remainders of 0 or 1 when divided by 3.

3. Perfect squares have only remainders of 0 or 1 when divided by 4.

4. Squares of prime numbers have exactly 3 divisors.

5. Consecutive perfect squares differ by an odd number:
$$n^2 - (n-1)^2 = n^2 - n^2 + 2n - 1 = 2n - 1$$

6. The prime factorization of a perfect square has only even exponents.

DIOPHANTINE EQUATIONS

Diophantine equations are equations with integer numbers. Most of the time, these equations have several unknowns but, the fact that they are integer provides additional information that we use to find the solutions.

Some of the facts used when solving Diophantine equations are:

- the parity of numbers (even or odd)

- the divisibility of numbers by various prime factors

- the fundamental theorem of arithmetic

- the last digit(s) of the numbers

In this workbook, we shall solve Diophantine equations based on factoring. At the next level, we shall study more general algorithms for solving linear Diophantine equations.

Diophantine equations solved by factoring rely on the fact that the prime factorization of a number is unique. In the following equation:

$$m \times n = 13 \times 17$$

the factors on both sides must match. One of the numbers m and n must be equal to 13 and the other to 17. Or, one of the numbers must be equal to 1 and the other to 221. If negative numbers are allowed as solutions, $(-1, -221)$ and $(-13, -17)$ are also in the solution set.

When solving such an equation, we must first bring it to a form with factors containing the unknowns on one side and a number on the other side. Factoring the number and applying the fundamental theorem of arithmetic are the next steps.

Example Solve the equation if k and p are positive integers:

$$3k + kp = 6p + 57$$

Step 1   Bring all the unknowns to one side:

$$3k + kp - 6p = 57$$

Step 2   Factor the side containing the unknowns. Like in this example, one may need to force a factor:

$$
\begin{aligned}
k(3 + p) - 6p - 18 &= 57 - 18 \\
k(3 + p) - 6(p + 3) &= 39 \\
(3 + p)(k - 6) &= 39
\end{aligned}
$$

Now factor the number on the right side and apply the uniqueness of the prime factorization:

$$(3 + p)(k - 6) = 3 \times 13$$

Summarize the possible solutions. Select only the ones that have positive values for p and k:

Case 1	$3 + p = 13$ $k - 6 = 3$	$p = 10$ $k = 9$	valid solution
Case 2	$3 + p = 39$ $k - 6 = 1$	$p = 36$ $k = 7$	valid solution

Negative choices for the two factors are possible, but they all lead to negative solutions.

PRACTICE SIX

Do not use a calculator for any of the problems!

Exercise 1

If m and k are non-zero integers, how many solutions does the equation have:

$$km + m = 6$$

(A) 1

(B) 2

(C) 3

(D) 7

Exercise 2

If m and k are integers, what is the largest possible value of $m \cdot k$:

$$km - m + k = 16$$

(A) 12

(B) 16

(C) 28

(D) 32

Exercise 3

What is the largest number of positive integers greater than 1 that have a product of 5400?

Exercise 4

Each side of a convex quadrilateral is labeled with a positive integer. Each vertex of the quadrilateral is labeled with a number that is the product of the numbers on the two sides that meet at the vertex. The numbers that label the vertices have a sum of 14. What is the sum of the numbers that label the sides?

(A) 4

(B) 7

(C) 9

(D) 15

Exercise 5

How many pairs of different positive integers have a sum of 67?

(A) 33

(B) 34

(C) 66

(D) 67

Exercise 6

The set M contains a number of distinct (i.e. different) odd integers. We form all possible positive differences among their squares. At least how many integers should there be in the set M so as to have at least two differences that end in the same digit?

(A) 3

(B) 4

(C) 5

(D) there is not enough information

Exercise 7

The set J contains fourth powers of N distinct integers. How many possible values are there for the last digit of the product of the numbers in J?

(A) 4

(B) 6

(C) 8

(D) 9

Exercise 8

The middle number among three consecutive integers is a perfect square. From the choices below, select the gcf of the products of all such sets of three numbers.

(A) 1

(B) 4

(C) 12

(D) 60

Exercise 9

Write 6^{2015} as a sum of three perfect cubes.

Exercise 10

In a $N \times M$ grid of white square grid cells, three adjacent cells that are either in the same row or in the same column are colored blue. If there are 416 ways to do this, how many grid cells does the grid have?

(A) 208

(B) 240

(C) 416

(D) 832

Exercise 11

Dina and Lila are playing cards. Dina has N cards. Lila has either 2 cards more or 2 cards less than Dina. If Dina does not know how many cards Lila is holding, then N is:

(A) 3 or more

(B) 3 or less

(C) a multiple of 4

(D) a multiple of $N + 2$

Exercise 12

What are the values of the integers k and p?

$$2^k + 3^p = 857$$

Exercise 13

In a storm relief medical facility, a certain amount of supplies is needed on average for each case. If there were 3 cases fewer per day, the 77 existing supply packs would last 5 days longer. How many cases are seen each day? If there are several solutions, list them all.

Exercise 14

If x and y have integer values, how many distinct solutions does the following equation have?

$$|x - y| + y^2 = 9$$

MISCELLANEOUS PRACTICE

Do not use a calculator for any of the problems!

Exercise 1

For how many positive integer values of k is the fraction an integer:

$$\frac{10 + k}{k - 1}$$

(A) 0

(B) 1

(C) 2

(D) 4

Exercise 2

When dividing a positive integer by 17, we obtain a quotient equal to the remainder. What is the largest digit sum the number can have?

(A) it can have any digit sum

(B) 17

(C) 18

(D) 27

Exercise 3

The number N is formed by writing the numbers from 1 to 100 in increasing order in a row:

$$N = 12345 \cdots 9899100$$

Remove 16 digits and leave the remaining digits put, so as to obtain the smallest possible number. What is the sum of the digits you erased?

Exercise 4

If the sum of 45 positive consecutive integers is a perfect square, what is the smallest possible value of the median number?

(A) 5

(B) 20

(C) 25

(D) 45

Exercise 5

Order the numbers by increasing value:

$$0.\overline{9374}, \ 0.9\overline{374}, \ 0.93\overline{74}, \ 0.937\overline{4}, \ 0.\overline{93}, \ 0.9\overline{43}$$

Exercise 6

How many pairs of positive integers have a sum of 305 and a remainder of 4 when the larger number is divided by the smaller one?

Exercise 7

Write 5^{2013} as a sum of two perfect squares.

Exercise 8

30 consecutive numbers have a positive sum smaller than 465. What is the product of the 30 consecutive numbers?

Exercise 9

30 consecutive numbers have a sum smaller than 465. What is the largest such sum?

Exercise 10

If P and Q are positive integers and:

$$(Q + 5) \cdot (P - 7) = 2 \cdot Q \cdot P$$

How many possible solutions are there?

Exercise 11

  i. What is the sum of all 3-digit positive integers with digits 1, 2 and 3, if each digit is used once in each number?

  ii. What is the sum of all 3-digit positive integers with digits 1, 2 and 3, if digits can be used repeatedly?

Exercise 12

What is the sum of all 5-digit positive integers with digits 1, 3, 5, 7, and 9, if each digit is used once in each number?

Exercise 13

What is the sum of all 4-digit numbers with 2 digits of 2 and 2 digits of 7?

Exercise 14

What is the sum of all the positive integers that have a digit sum and a digit product both equal to 14?

Exercise 15

What is the sum of the first 25 decimal digits of each of the following numbers:

i. $\sqrt{0.\underbrace{99\cdots9}_{25 \text{ digits}}}$

ii. $\sqrt{0.\underbrace{11\cdots1}_{25 \text{ digits}}}$

iii. $\sqrt{0.\underbrace{44\cdots4}_{25 \text{ digits}}}$

Exercise 16

In base 5, how many positive integers smaller than 1000 are multiples of 4?

Exercise 17

If b is an integer, the expression $b^2 + b + 1$ is:

(A) always even

(B) always odd

(C) sometimes even

Exercise 18

For how many integer values of t is the fraction below reducible?

$$\frac{t^2 - t + 4}{t^2 + t}$$

(A) 2

(B) 5

(C) 8

(D) infinity

Exercise 19

How many of the following fractions are irreducible?

$$\frac{589}{2}, \frac{589}{3}, \cdots \frac{589}{588}, \frac{589}{589}$$

Exercise 20

x, y, and z are unknown digits in the base-10 number $4xy77z$. Find the digits x, y, and z that make the number divisible by 888.

Exercise 21

How many distinct pairs of positive numbers have an lcm of 25?

(A) 1

(B) 2

(C) 3

(D) 4

Exercise 22

How many distinct pairs of positive numbers have an lcm of 36?

(A) 9

(B) 12

(C) 13

(D) 25

Exercise 23

It took Amon, the elephant, several days to eat all the melons in Sarawut's patch. If Amon had eaten 20 more melons per day during twice the number of days, he would have eaten 265 more melons. How many melons did Amon eat each day, on average?

Exercise 24

What is the largest possible digit sum of the digit sum of a 4-digit number?

Exercise 25

What is the largest possible digit product of the digit sum of a 4-digit number?

Exercise 26

The following numbers are in base 6. Which one of them is divisible by 5?

(A) 1234

(B) 1235

(C) 1254

(D) 1255

Exercise 27

If $a, b, c,$ and m are base-10 digits, which of the following must be a common factor of the numbers $abcabc$ and $m77m$?

(A) a

(B) $a + m$

(C) $7a$

(D) 5

(E) 77

Exercise 28

How many digits of 5 must the following number have in order to be a perfect square:

$$1\underbrace{555\ldots55}_{k \text{ digits}}$$

Exercise 29

Three consecutive positive integers have an lcm of 976500. What is the smallest possible product of the three numbers?

Exercise 30

In base 2, a number is divisible by 3 if:

(A) its digit sum is a multiple of 3.

(B) its last two digits are 11.

(C) its alternating digit sum is a multiple of 3.

(D) the number of its digits is divisible by 3.

SOLUTIONS TO PRACTICE ONE

Do not use a calculator for any of the problems!

Solution 1

Do mental checks on the divisibility of the numerators and denominators. Figure out if/how the fraction simplifies. Check what kind of factors are left at the denominator after simplification. The following considerations should be made mentally or with very little computation:

$\dfrac{11}{8}$	T	The denominator only has factors of 2.
$\dfrac{123}{144}$	NTM	Considering divisors that are powers of 3, we find that the denominator is divisible by 9 and the numerator is divisible only by 3. A factor of 3 remains at the denominator after simplification. The numerator is odd, the denominator is even - there are factors of 2 at the denominator after simplification.

$\dfrac{18}{144}$	T	The denominator is divisible by 9 and the numerator is divisible by 9. All the other factors remaining at the denominator are factors of 2. The equivalent decimal is terminating.
$\dfrac{3}{256}$	T	The denominator only has factors of 2 (recognize 256 as a power of 2!).
$\dfrac{121}{275}$	T	Both the numerator and the denominator are divisible by 11. After simplification, the denominator has only factors of 5.
$\dfrac{11}{36}$	NTM	The denominator has factors of 3, as well as factors of 2.
$\dfrac{7}{49}$	NTR	The denominator has a factor of 7 after simplification and no factors of 2 or of 5.
$\dfrac{693}{132}$	T	The denominator is divisible by 3 and 11. The numerator is also divisible by 3 and by 11. After simplification, only factors of 2 will remain at the denominator (since $132 = 11 \times 3 \times 4$).
$\dfrac{5}{14}$	NTM	The denominator has a factor of 2 and a factor of 7.

Exercise 2

Sort in decreasing order:

$$a = 6.538, \ b = 6.5\overline{38}, \ c = 6.\overline{538}, \ d = 6.53\overline{7}$$

Solution 2

The correct answer is (C): $c > b > a > d$.

Exercise 3

If k is a positive integer, for which value of k is the fraction below equivalent to the largest possible non-terminating repeating decimal:

$$\frac{14}{7 \times \heartsuit}$$

Solution 3

The fraction simplifies by 7. To obtain the largest value, we need the smallest possible denominator. The smallest denominator that makes the equivalent decimal non-terminating is 1. We take advantage of the fact that *any integer can be written as a non-terminating repeating decimal*:

$$\frac{14}{7} = 1.\overline{9}$$

The correct answer is (A).

Note: More generally, any terminating decimal can be written as a non-terminating decimal.

Exercise 4

k, q, p, and m are positive integer values that make the fraction below equivalent to the largest possible proper, non-integer, non-terminating repeating decimal:

$$\frac{2^4 \times 5^3 \times 7^2 \times 3^5}{2^k \times 5^q \times 7^p \times 3^m}$$

Find $k + q + p + m$.

Solution 4

For the decimal to not be mixed, there should be no factors of 2 or 5 at the denominator. Therefore, $k = 4$ and $q = 3$.

For the fraction to be the largest, the denominator must be the smallest possible. 1 cannot be used because the statement asks for non-integer values. We also have to ensure the fraction is proper. This is achieved by simplifying all the factors of 7 and keeping only one factor of 3. The fraction can be written as:

$$\frac{1}{2^{k-4} \times 5^{q-3} \times 7^{p-2} \times 3^{m-5}}$$

and it is equal to:

$$\frac{1}{3}$$

if and only if: $k = 4$, $q = 3$, $p = 2$, $m = 6$. Therefore,

$$k + q + p + m = 4 + 3 + 2 + 6 = 15$$

Exercise 5

Convert the following decimals into irreducible fractions:

Solution 5

Apply the method for converting repeating decimals and simplify each fraction by factoring the numerator and the denominator into primes.

$0.\overline{14}$	$1.\overline{5}$	$0.2\overline{34}$	$0.\overline{7896}$	$0.34\overline{5}$	$0.451\overline{3}$	$0.91\overline{28}$
$\dfrac{14}{99}$	$\dfrac{14}{9}$	$\dfrac{116}{495}$	$\dfrac{7889}{9990}$	$\dfrac{311}{900}$	$\dfrac{677}{1500}$	$\dfrac{9037}{9900}$

Exercise 6

Find the last non-zero decimal of the division:

$$\frac{17}{5^{2013}}$$

Solution 6

Since the denominator is a multiple of 5, the equivalent decimal is terminating. To find the last digit, transform the fraction so the denominator is a power of 10:

$$\frac{17}{5^{2013}} \times \frac{2^{2013}}{2^{2013}} = \frac{17 \times 2^{2013}}{10^{2013}}$$

Since the division by a power of 10 merely moves the decimal point to the left, the last digit of the decimal is the same as the last digit of the integer:

$$17 \times 2^{2013}$$

Since the powers of 2 end in the sequence 2, 4, 8, 6, 2, ..., the last digit of the power 2^{2013} is obtained by dividing the exponent into groups of 4:

$$2013 = 4 \times 503 + 1$$

Since the remainder is 1, the power of 2 ends in 2. The final digit of the product is the last digit of 7×2, which is 4.

Exercise 7

For which positive integers k is the value of the fraction:

$$\frac{2013}{2^k}$$

a decimal that ends in 5?

(A) only $k = 1$

(B) any k that is a multiple of 4 plus 1

(C) any k that is a multiple of 4 plus 1 or plus 3

(D) any k

Solution 7

Since the denominator is a multiple of 2, the equivalent decimal is terminating. Transform the fraction so that the denominator is a power of 10:

$$\frac{2013}{2^k} \times \frac{5^k}{5^k} = \frac{2013 \times 5^k}{10^k}$$

All powers of 5 end in 5. Since 2013 is odd, its product with a power of 5 ends in 5. The correct answer choice is (D).

Exercise 8

When converting the following fraction to a decimal, which digit is in the 64th decimal place?

$$\frac{1}{7}$$

Solution 8

The decimal must be non-terminating repeating. Perform long division to find the repetend of the decimal number:

$$\frac{1}{7} = 0.\overline{142857}$$

The repetend has 6 digits. Divide 64 by 6:

$$64 = 6 \times 10 + 4$$

The 64th digit after the decimal point is 8.

Exercise 9

Calculate the value of the expression:

Solution 9

$$
\cfrac{1}{3-\cfrac{1}{1-\cfrac{1}{1+0.\overline{6}}}} \;=\; \cfrac{1}{3-\cfrac{1}{1-\cfrac{1}{1+\cfrac{2}{3}}}}
$$

$$
=\; \cfrac{1}{3-\cfrac{1}{1-\cfrac{1}{\cfrac{5}{3}}}}
$$

$$
=\; \cfrac{1}{3-\cfrac{1}{1-\cfrac{3}{5}}}
$$

$$
=\; \cfrac{1}{3-\cfrac{1}{\cfrac{2}{5}}}
$$

$$
=\; \cfrac{1}{3-\cfrac{5}{2}}
$$

$$
=\; \cfrac{1}{\cfrac{1}{2}}
$$

$$
=\; 2
$$

Exercise 10

Find the digit k if it is true that:

$$5.\overline{k} = 6$$

Solution 10

$k = 9$. Note that:

$$0.\overline{9} = 1$$

Exercise 11

What is the value of the integer m?

$$\frac{1}{0.\overline{63}} = \frac{m+4}{m}$$

Solution 11

Convert the decimal into a fraction and solve for m:

$$
\begin{aligned}
\frac{1}{0.\overline{63}} &= \frac{1}{\dfrac{63}{99}} \\[2mm]
&= \frac{99}{63} \\[2mm]
&= \frac{11}{7} \\[2mm]
&= \frac{7+4}{7}
\end{aligned}
$$

$m = 7$

Exercise 12

There are integer values m, a, b, c, and d so that:

$$\frac{67}{155} = m + \cfrac{1}{a + \cfrac{1}{b + \cfrac{1}{c + \cfrac{1}{d}}}}$$

What is the value of $m + a + b + c + d$?

Solution 12

The value of m is zero, since the value of the fraction is smaller than 1.

At each step, take the integers out of the fraction and turn the remaining fraction into an improper fraction. Repeat the process until the last fraction becomes an integer:

$$\frac{67}{155} = \frac{1}{\dfrac{155}{67}}$$

$$= \frac{1}{2 + \dfrac{21}{67}}$$

$$= \frac{1}{2 + \dfrac{1}{\dfrac{67}{21}}}$$

$$= \frac{1}{2 + \dfrac{1}{3 + \dfrac{4}{21}}}$$

$$= \frac{1}{2 + \dfrac{1}{3 + \dfrac{1}{\dfrac{21}{4}}}}$$

$$= \frac{1}{2 + \dfrac{1}{3 + \dfrac{1}{5 + \dfrac{1}{4}}}}$$

The integers are 0, 2, 3, 4, and 5. Their sum is 14.

Note: The numbers a, b, c, d are called *partial quotients*.

Exercise 13

If k is a 2-digit number, what percentage of the fractions:

$$\frac{1}{k}$$

are equivalent to terminating decimals?

Solution 13

There are 90 2-digit numbers in total. We have to count how many of them have factors of 2 or of 5 only. There are 3 numbers that are powers of 2: $16, 32$, and 64. There is one number that is a power of 5: 25. There is only one cross-product that is smaller than 100: $5 \times 16 = 80$. However, there are also two digit numbers that are products of 1-digit numbers: $5 \times 2 = 10, 5 \times 4 = 20, 5 \times 8 = 40, 25 \times 2 = 50$. In total, there are 9 numbers that satisfy. The percentage is:

$$\frac{9}{90} = 10\%$$

Exercise 14

A rectangle has side lengths 19 and 47. Lila cuts it to obtain the largest possible square and another rectangle. She then continues to cut the remaining rectangle in the same way until there is no rectangle left. How many squares does she have at the end?

Solution 14

Notice how this process is the same as writing the continued fraction:

$$
\begin{aligned}
\frac{47}{19} &= 2 + \frac{9}{19} \\[2ex]
&= 2 + \cfrac{1}{\cfrac{19}{9}} \\[2ex]
&= 2 + \cfrac{1}{2 + \cfrac{1}{9}}
\end{aligned}
$$

There will be a total of 13 squares:

2 19 × 19 squares

2 9 × 9 squares

9 1 × 1 squares

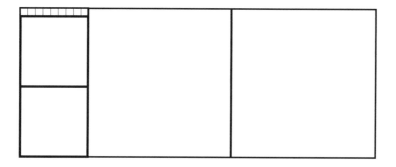

Exercise 15

The number 0.2981736451 . . .:

(A) is rational

(B) is irrational

(C) could be either rational or irrational

(D) is neither rational nor irrational, since it is decimal

Solution 15

The correct answer is (C). While irrational numbers have an infinite number of decimals that do not follow a repeating pattern, it is not clear from the fragment shown that the number is irrational. Some rational numbers have very large periods. Indeed, it is extremely easy to build a rational number that has a repetend identical to the decimals shown:

$$\frac{2981736451}{9999999999}$$

Unless more information is provided, we cannot be sure about the nature of this number.

Solutions to Practice Two

Do not use a calculator for any of the problems!

Exercise 1

Dividing a 2-digit positive integer by 17 produces a remainder of 9. What is the difference between the smallest such integer and the largest such integer?

Solution 1

The smallest quotient is 1. The smallest integer with the property is:

$$17 \times 1 + 9 = 26$$

The largest multiple of 17 that has 2 digits is 85. Therefore, the largest number with the property is:

$$17 \times 5 + 9 = 85 + 9 = 94$$

The difference is: $94 - 26 = 68$.

Exercise 2

What is the sum of the integer numbers that produce a quotient of 19 when divided by 11?

Solution 2

There are 11 possible remainders. The numbers are:

$$19 \times 11 + 0$$

$$19 \times 11 + 1$$

$$\ldots$$

$$19 \times 11 + 10$$

The sum of these numbers is:

$$
\begin{aligned}
S &= 19 \times 11 \times 11 + 1 + 2 + \cdots + 10 \\
&= 19 \times 121 + \frac{10 \times 11}{2} \\
&= 20 \times 121 - 121 + 55 \\
&= 2420 - 66 = 2354
\end{aligned}
$$

Exercise 3

Which of the following choices is the irreducible form of the fraction:

$$\frac{12155}{17017}$$

(A) $\dfrac{5}{7}$

(B) $\dfrac{1105}{1547}$

(C) $\dfrac{85}{119}$

(D) $\dfrac{12155}{17017}$

Solution 3

We factor the numerator and the denominator into primes:

$$\frac{12155}{17017} = \frac{5 \times 11 \times 13 \times 17}{7 \times 11 \times 13 \times 17}$$

$$= \frac{5}{7}$$

The correct answer choice is (A).

Exercise 4

If the product of three consecutive numbers has a factor of 3^6, what is the smallest possible value for the smallest of the three numbers?

(A) 3^2

(B) 3^3

(C) $3^2 - 2$

(D) $3^6 - 2$

Solution 4

Among three consecutive numbers only one of them is a multiple of 3. Therefore, only one of them can be a multiple of 3^6. Its smallest possible value is 3^6. By Setting 3^6 as the largest of the three consecutive numbers, we find that the smallest of the three numbers must be $3^6 - 2$.

The correct answer choice is (D).

Exercise 5

If the product of three consecutive numbers has a factor of 2^9, what is the smallest possible value for the smallest of the three numbers?

Solution 5

Among three consecutive numbers there are either one or two multiples of 2. Distributing the nine factors of 2 among the three consecutive numbers can be done only in the following ways:

- $2 \times m, 2 \times m + 1, 2^8 \times p$, where $2^8 \times p = 2 \times m + 2$;

- $2k + 1, 2^9 \times p, 2k + 3$, where $2^9 \times p = 2k + 2$;
- $2^8 \times p, 2^8 \times p + 1, 2(2^7 \times p + 1)$.

It is important to notice that there must be either one number that has all the factors of 2, or, if split among two different numbers, the factors of 2 can only be split as one factor and the remaining 8 factors. This is because more than one multiple of 4, 8, etc. cannot be found within three consecutive numbers.

Now, to find the smallest number possible, we make all other factors equal to 1. The sequences of three consecutive numbers become:

- $2^8, 2^8 + 1, 2^8 + 2$
- $2^9 - 1, 2^9, 2^9 + 1$
- $2^8 - 2, 2^8 - 1, 2^8$

Of these, the last sequence has the smallest value of the smallest term: $2^8 - 2$.

Exercise 6

Factor into primes:

Solution 6

$$
\begin{aligned}
847 &= 7 \times 11^2 \\
8189 &= 19 \times 431 \\
6754 &= 2 \times 11 \times 307 \\
4345 &= 5 \times 11 \times 79 \\
2132 &= 2^2 \times 13 \times 41 \\
4947 &= 3 \times 17 \times 97
\end{aligned}
$$

Exercise 7

The positive difference (*positive difference* means that we subtract the smaller number from the larger number) of the squares of two consecutive numbers is:

(**A**) always odd

(**B**) always even

(**C**) sometimes odd

Solution 7

The positive difference of the squares of two consecutive numbers is *always odd.* Denote the numbers by k and $k+1$. Subtract their squares:

$$(k+1)^2 - k^2 = k^2 + 2k + 1 - k^2 = 2k + 1$$

Exercise 8

Find three prime numbers a, b, and c such that:

$$a + b + c = 516$$

Solution 8

Since the sum is even, not all three numbers can be odd. Since they are all prime, only one of them can be even and must be equal to 2. The remaining two numbers must be odd and have a sum of 514.

Subtract odd primes from 514 in increasing order until you find a difference that is prime:

$$514 - 3 = 511 = 7 \times 73$$
$$514 - 5 = 509$$

509 is prime. A possible solution is: 2, 5, and 509. Note that there are other solutions possible such as 2 11, and 503 but the statement only asked to find one possible solution.

Exercise 9

Find the prime numbers k and p such that:

$$2^k + p = 151$$

Solution 9

Since 151 is not large, it is easiest to solve this problem by brute force. Powers of 2 with prime exponents that do not exceed 151 are: 2^2, 2^3, 2^5, 2^7:

$$151 = 4 + 147$$
$$151 = 8 + 143$$
$$151 = 32 + 119$$
$$151 = 128 + 23$$

Of 147, 143, 119, and 23 only 23 is prime. Therefore, $k = 7$ and $p = 23$.

Exercise 10

Two consecutive numbers p and k are prime. Find:

$$k^p + p^k$$

Solution 10

Two consecutive numbers have different parity: one is even and one is odd. Since the only even prime is 2, the numbers must be 2 and 3. We calculate:

$$2^3 + 3^2 = 17$$

Exercise 11

Write 3^5 as a sum of three consecutive numbers. What is the largest prime divisor among the prime divisors of the three numbers?

(A) 1

(B) 2

(C) 3

(D) 41

Solution 11

The correct answer is (D).

Let the consecutive numbers be: $n - 1$, n, $n + 1$:

$$n - 1 + n + n + 1 = 3^5$$

$$3 \times n = 3^5$$

$$n = 3^4 = 81$$

The consecutive numbers are, with their prime factorings:

$$80 = 2^4 \times 5$$

$$81 = 3^4$$

$$82 = 2 \times 41$$

The largest prime divisor among all divisors is 41.

Exercise 12

Write 2^7 as a sum of consecutive odd numbers.

Solution 12

Split the power of two into a sum of two odd terms, as follows:

$$
\begin{aligned}
2^7 &= 2 \times 2^6 \\
&= 2^6 + 2^6 \\
&= 2^6 + 1 + 2^6 - 1
\end{aligned}
$$

The numbers are: 63 and 65.

Therefore, *it is always possible to obtain 2^k (k $\geq$ 2)as a sum of two consecutive odd numbers.*

Exercise 13

The remainder from dividing N by 34 is 31. What is the remainder from dividing N by 17?

Solution 13

Write the integer division as:

$$N = 34 \times q + 31$$

which is also:

$$N = 17 \times 2 \times q + 31$$

The remainder cannot be larger than the divisor, so we must use:

$$31 = 17 + 14$$

and write:

$$N = 17 \times 2 \times q + 17 + 14 = 17 \times (2q + 1) + 14$$

The remainder is 14.

Exercise 14

Write the number 33306 as a product of two consecutive numbers.

Solution 14

Factor into primes:

$$33306 = 2 \times 3 \times 7 \times 13 \times 61$$

Consider the fact that the two consecutive factors must have last digits that are consecutive. Therefore, we notice that $2 \times 3 \times 3$ ends in 8 and 7×1 ends in seven. Regroup the primes to form two consecutive factors:

$$3 \times 61 = 183$$
$$2 \times 7 \times 13 = 182$$

Hint for another solution path: Notice that the two consecutive factors must be close in value to an approximation of the square root of 33306. Since $18 \times 18 = 324$, it is useful to look at factors that are slightly larger than 180.

SOLUTIONS TO PRACTICE THREE

Do not use a calculator for any of the problems!

Exercise 1
List the proper divisors of 531.

Solution 1
First, factor the number into primes:

$$531 = 3^2 \times 59$$

List the proper divisors:

$$1, \ 3, \ 9, \ 59, \ 177$$

Exercise 2
What is the largest odd divisor of 504?

Solution 2
Factor 504 into primes:

$$504 = 2^3 \cdot 3^2 \cdot 7$$

and select the odd factors. The largest odd divisor is 63.

Exercise 3

How many positive integers smaller than 1000 have exactly 3 proper divisors?

(A) 1

(B) 2

(C) 3

(D) more than 3

Solution 3

The number of positive divisors of a number N with a generic prime factorization:

$$N = p_1^{a_1} \times p_2^{a_2} \times \cdots p_k^{a_k}$$

is equal to:

$$\tau(N) = (a_1 + 1)(a_2 + 1) \cdots (a_k + 1)$$

(This theorem is explained in the *Practice Combinatorics, level 3* volume of this series.)

The only numbers that have exactly 4 proper divisors are numbers that factor into primes in the form $N = p^4$, where p is a prime. The set of proper divisors is:

$$1, \ p, \ p^2, \ p^3$$

The numbers smaller than 1000 that satisfy this requirement are:

$$2^4 \ = \ 16$$

$$3^4 \ = \ 81$$

$$5^4 \ = \ 625$$

There are three such numbers. The correct answer is (C).

Exercise 4

What is the remainder if we divide 540 by the sum of its positive divisors?

Solution 4

The sum of the positive divisors is certainly larger than 540 since 540 is a positive divisor of itself. Therefore, the quotient of such a division is zero and the remainder is 540.

Exercise 5

The number 2013 can be written as a sum of k consecutive integers. How many of the following numbers are valid values for k?

$$11, \ 29, \ 33, \ 57, \ 61, \ 73, \ 99, \ 183, \ 671$$

Solution 5

Denoting the first number by $m + 1$ and the last number by $m + k$, so that there is a total of k consecutive numbers, the sum of the numbers must be:

$$
\begin{aligned}
S &= m + 1 + m + 2 + m + 3 + \cdots + m + k \\
&= k \cdot m + 1 + 2 + 3 + \cdots + k \\
&= k \cdot m + \frac{k(k+1)}{2}
\end{aligned}
$$

The Diophantine equation is:

$$
\begin{aligned}
k \cdot m + \frac{k(k+1)}{2} &= 2013 \\
2km + k(k+1) &= 4026 \\
k(2m + k + 1) &= 2 \cdot 3 \cdot 11 \cdot 61
\end{aligned}
$$

k is an element of the set of divisors of 4026:

$$1, \ 2, \ 3, \ 6, \ 11, \ 22, \ 33, \ 61, \ 66, \ 122, \ 183, \ 366, \ 671, \ 1342, \ 2013, \ 4026$$

However, since $k < 2m + k + 1$, only the first half of the 16 divisors can be values of k. Moreover, $k = 1$ does not yield consecutive integers, since it results in a list of one number. So k can be any of:

$$2, \ 3, \ 6, \ 11, \ 22, \ 33, \ 61$$

The correct answer is (A).

Exercise 6

If n is a positive integer greater than 1, what are the known numeric divisors of the expression:

$$(n - 1)n(n + 1)$$

Solution 6

The factors $n - 1$, n, and $n + 1$ are consecutive integers. Among them, there must be at least one multiple of 2 and exactly one multiple of 3.

The expression has the divisors: 1, 2, 3, and 6, as well as possibly other divisors that cannot be known without knowing the value of n.

Exercise 7

If n is a positive integer, and the largest power of 2 that divides the expression $E = (n - 1)n(n + 1)$ is 2^{12}, how many of the following statements could be true?

   i. Each of $n - 1$, n, and $n + 1$ must be divisible by 8.

   ii. Each of $n - 1$, n, and $n + 1$ must be divisible by 16.

   iii. n is divisible by 2^{12}.

   iv. $n - 1$ is divisible by 2^{12}.

   v. Either one of $n - 1$ or $n + 1$ is divisible by 2^{11}, while the other must be divisible by 2.

Solution 7

Consecutive multiples of 8 or 16 are 8 and 16 numbers apart on the number line, respectively. Consecutive integers such as $n - 1$, n, and

$n + 1$ cannot all be multiples of 8 or 16. (**i.**) and (**ii.**) are false.

(**iii.**) can be true, if n is even.

(**iv.**) cannot be true. If $n - 1$ is even, then $n + 1$ is even as well. The expression would be divisible by 2^{13}, contradicting the statement.

(**v.**) can be true if n is odd.

Therefore, two of the statements could be true.

Exercise 8

When dividing k positive consecutive numbers by 5, the sum of the remainders is 31. What is the product of the remainders?

(A) 0

(A) 1

(A) 2

(A) 24

Solution 8

When dividing consecutive positive numbers by 5, the remainders form the sequence:

$$0, \ 1, \ 2, \ 3, \ 4, \ 0, \ 1, \ \cdots$$

For the sum of the remainders to be 31 we need to have more than 5 consecutive numbers. Since there is a remainder of 0 in any one subsequence of 5 remainders, the product must be 0.

The correct answer is (A).

Exercise 9

If a and b are randomly chosen non-zero digits, which of the numbers below is a divisor of *baab*?

(A) 3

(B) 7

(C) 11

(D) 13

Solution 9

The correct answer is (C). In expanded form, the number *baab* can be written as:

$$b \times 1001 + a \times 110$$

Factor the numbers in the expression:

$$b \times 7 \times 11 \times 13 + a \times 2 \times 5 \times 11$$

Only 11 is a common factor:

$$11 \times (b \times 7 \times 13 + a \times 11)$$

Exercise 10

What is the sum $a + b + c$ if a, b, and c are different digits that make the number *7a5bc* divisible by 2, by 5, and by 11?

Solution 10

For the number to be divisible by 2 and by 5 the digit c must be zero.

For the number *7a5b0* to be divisible by 11 the alternating digit sum must be zero or a multiple of 11. If:

$$0 - b + 5 - a + 7 = 0$$

we have $a+b = 12$ and $a+b+c = 12$. The correct answer choice is (C).

An alternate digit sum that is a non-zero multiple of 11 is not possible, since two digits cannot add up to more than 18.

Exercise 11

What is the remainder of the integer division $a \div b$ if:

$$a = 1 + 2 + 3 + \cdots + 90$$

and

$$b = 2 + 4 + 6 + \cdots + 12$$

Solution 11

$$a = \frac{90 \cdot 91}{2} = 45 \cdot 91 = 3^2 \cdot 5 \cdot 7 \cdot 13$$

$$b = 2(1 + 2 + 3 + \cdots + 6) = 2 \cdot \frac{6 \cdot 7}{2} = 2 \cdot 3 \cdot 7$$

Write the integer division:

$$
\begin{aligned}
a &= bq + r \\
45 \cdot 91 &= 6 \cdot 7 \cdot q + r \\
21 \cdot 15 \cdot 13 &= 21 \cdot 2 \cdot q + r \\
21 \cdot 15 \cdot 13 &= 21 \cdot 2 \cdot q + 21 \cdot p \\
15 \cdot 13 &= 2 \cdot q + p
\end{aligned}
$$

Since dividing an odd number by 2 must give the remainder $p = 1$, we have that $r = 21$.

Note: This problem was assigned in order to show why it is not a good idea to simplify the fraction $\frac{a}{b}$ by 21 if we don't plan to multiply the 21 back in once we find the remainder. It's a common error that on such a problem the student will answer that the remainder is 1.

Exercise 12

The 2-digit integer ab has different digits a and b such that a is twice b. What is the sum of all such 2-digit numbers?

Solution 12

Write the number in expanded form:

$$ab = 10 \times a + b$$

and use the fact that $a = 2b$:

$$ab = 10 \times 2 \times b + b = 21 \times b$$

Since a is a non-zero digit, b can only have the values 1, 2, 3, and 4. The sum of all such numbers is equal to 210:

$$
\begin{aligned}
S &= 21 \times (1 + 2 + 3 + 4) \\
&= 21 \times \frac{4 \times 5}{2} \\
&= 21 \times 10 = 210
\end{aligned}
$$

Exercise 13

Which of the following are divisors of any repdigit with 5 digits?

(A) 5

(B) 11

(C) 41

(D) 101

Solution 13

A 5-digit repdigit factors as follows:

$$ddddd = d \times 11111 = d \times 41 \times 271$$

Only choice (C) is valid.

Exercise 14

If a, b, and c are digits, how many of the following numbers must be divisors of $abcabcabc$?

$$abc, \ 1, \ 111, \ 1001001, \ 3, \ 333667, \ a00a00a00, \ bc, \ 9, \ 1001$$

(A) 1

(B) 3

(C) 4

(D) 5

Solution 14

abc is a divisor because:

$$abc \times 1001001 = abcabcabc$$

1001001 is divisible by 3 since it has a digit sum of 3.

Dividing 1001001 by 3 we find that:

$$1001001 = 3 \times 333667$$

Therefore, 333667 is also a divisor.

The other numbers are not necessarily divisors. The correct answer is (C).

Exercise 15

What is the largest 5-digit palindrome divisible

   i. by 8

   ii. by 5

   iii. by 11

Solution 15

   i. To be divisible by 8 the palindrome must end in a 3 digit number that is divisible by 8. This number has to end in the largest possible digit, since this will also be the first digit of the palindrome: 888. The palindrome is 88888.

ii. To be divisible by 5 the palindrome must end in 5. The remaining digits may be as large as possible. The palindrome is 59995.

iii. To be divisible by 11 the alternating digit sum must be a multiple of 11. Assume the palindrome looks like this: *abcba*. Then the digits must satisfy:

$$a - b + c - b + a \ = \ 11 \cdot k$$
$$2a - 2b + c \ = \ 11 \cdot k$$

Consider that digits have a maximum value of 9. Therefore, the alternate digit sum $2a - 2b + c$ must be one of 0, 11, or 22, as the largest possible value of $2a - 2b + c$ is 27 and is attained for $a = c = 9$ and $b = 0$.

We have to attempt to make a as large as possible. If we make $a = 9$, it is possible to make the alternate digit sum equal to zero by choosing $b = 9$ and $c = 0$. This yields the largest palindrome with the desired property:

$$99099$$

Exercise 16

An integer has 111 digits of 4 and 111 digits of 7. Is there an arrangement of the digits that makes this number a perfect square?

Solution 16

The digit sum of such a number is:

$$S = 4 \cdot 111 + 7 \cdot 111 = 11 \cdot 111 = 1221$$

We can easily check that S is divisible by 3 but not by 9. In a perfect square, all prime factors must occur with even exponents. Since the digit sum is the same, there is no arrangement of the digits that can make the number a perfect square.

Exercise 17

If n is a non-negative integer, the number of zeros at the end of $n!$ forms the sequence:

$$0, \ 0, \ 0, \ 0, \ 0, \ 1, \ 1, \ 1, \ 1, \ 1, \ 2, \ 2, \ 2, \ldots$$

The 26th term of this sequence is:

(A) 3

(B) 4

(C) 5

(D) 6

Solution 17

The correct answer is (D). The terms of the sequence increase by 1 each time a multiple of 5 is encountered. However, upon reaching 25!, there are 2 additional factors of five. The term of rank 26 is 25! and ends in 6 zeros.

Exercise 18

The 3-digit number A produces a remainder of 3 when divided by 5, and a remainder of 6 when divided by 8. What is the smallest possible value of A?

Solution 18

Write the division theorem for both cases:

$$A \ = \ 5k + 3$$

$$A \ = \ 8m + 6$$

Notice that $5 - 3 = 2$ and $8 - 6 = 2$ as well. Then:

$$A + 2 \ = \ 5k + 5 = 5(k + 1)$$

$$A + 2 \ = \ 8m + 8 = 8(m + 1)$$

and $A + 2$ is a multiple of both 5 and 8. Since 5 and 8 are coprime, $A + 2$ must be divisible by 40. The smallest 3-digit number divisible by 40 is 120. Therefore, $A = 118$.

Exercise 19

Test the method of casting out 8s for 3-digit numbers on each of the following numbers and circle the ones that are divisible by 8:

$$224, \ 136, \ 482, \ 864, \ 942, \ 748$$

Solution 19

Calculate the value of the test expressions:

$$22 + \frac{4}{2} \ = \ 24 \quad 224 \text{ is divisible by 8}$$

$$13 + \frac{6}{2} \ = \ 16 \quad 136 \text{ is divisible by 8}$$

$$48 + \frac{2}{2} \ = \ 49 \quad 482 \text{ is not divisible by 8}$$

$$86 + \frac{4}{2} \ = \ 88 \quad 864 \text{ is divisible by 8}$$

$$94 + \frac{2}{2} \ = \ 85 \quad 942 \text{ is not divisible by 8}$$

$$74 + \frac{8}{2} \ = \ 78 \quad 748 \text{ is not divisible by 8}$$

Exercise 20

Find the digits a, b, and c which make the number $3a4b5c$ a multiple of 72.

Solution 20

For a number to be a multiple of 72 it must be a multiple of 8 and a multiple of 9 (since 8 and 9 are coprime.)

For a number to be a multiple of 9, its digit sum must be a multiple of 9:

$$4 + b + 5 + c \ = \ 9 \cdot k$$

$$b + c \ = \ 9 \cdot m$$

Since the sum of two digits cannot exceed 18, the only possible values for $b + c$ are:

$$b + c = 9$$
$$b + c = 18$$

Since a sum of 18 is achieved only if both digits are 9, while $4b5c$ must be even, makes this choice impossible. Therefore, $b + c = 9$. With c even:

$$c = 0, b = 9$$
$$c = 2, b = 7$$
$$c = 4, b = 5$$
$$c = 6, b = 3$$
$$c = 8, b = 1$$

For the number to be divisible by 8 $b5c$ has to be divisible by 8. Since only 52 and 56 are divisible by 4, the only numbers to try are: 752 and 356. Of these, only 752 is divisible by 8.

Solutions to Practice Four

> Do not use a calculator for any of the problems!

Exercise 1

Convert the following decimal numbers to base 8 (octal):

$$56, \ 89, \ 105, \ 156, \ 203, \ 304, \ 392$$

Solution 1

$$56_{10} = 7 \times 8 = 70_8$$

$$89_{10} = 1 \times 8^2 + 3 \times 8 + 1 = 131_8$$

$$105 = 1 \times 64 + 5 \times 8 + 1 = 151_8$$

$$156 = 2 \times 64 + 3 \times 8 + 4 = 234_8$$

$$203 = 3 \times 64 + 1 \times 8 + 3 = 313_8$$

$$304 = 4 \times 64 + 6 \times 8 = 460_8$$

$$392 = 6 \times 64 + 1 \times 8 = 610_8$$

Exercise 2

Convert the following decimal numbers to base 16 (hexadecimal):

$$41, \ 92, \ 107, \ 203, \ 410, \ 548, \ 619$$

Solution 2

In base 16, the following digits are used in addition to the base-10 digits: $A = 10$, $B = 11$, $C = 12$, $D = 13$, $E = 14$, and $F = 15$:

$$41 \ = \ 2 \times 16 + 9 = 29_{16}$$

$$92 \ = \ 5 \times 16 + 12 = 5C_{16}$$

$$107 \ = \ 6 \times 16 + 11 = 6B_{16}$$

$$203 \ = \ 12 \times 16 + 11 = CB_{16}$$

$$410 \ = \ 1 \times 16^2 + 9 \times 16 + 10 = 19A_{16}$$

$$548 \ = \ 2 \times 16^2 + 2 \times 16 + 4 = 224_{16}$$

$$619 \ = \ 2 \times 16^2 + 6 \times 16 + 11 = 26B_{16}$$

Exercise 3

Convert the following decimal numbers to base 2 (binary):

$$14, \ 21, \ 25, \ 30, \ 65, \ 78, \ 91$$

Solution 3

$$14 \ = \ 2^3 + 2^2 + 2 = 1110_2$$

$$21 \ = \ 2^4 + 2^2 + 1 = 10101_2$$

$$25 \ = \ 2^4 + 2^3 + 1 = 11001_2$$

$$30 \ = \ 2^4 + 2^3 + 2^2 + 2 = 11110_2$$

$$65 \ = \ 2^6 + 1 = 1000001_2$$

$$78 \ = \ 2^6 + 2^3 + 2^2 + 2 = 1001110_2$$

$$91 \ = \ 2^6 + 2^4 + 2^3 + 2 + 1 = 1011011_2$$

Exercise 4

Convert the following numbers to base 10 (decimal):

$$54_8, \quad 88_9, \quad A1_{16}, \quad BBA_{13}, \quad 101_3, \quad 214_6, \quad 337_8$$

Solution 4

Simply use the expanded form of the number in the respective base. (Base 10 will be omitted from the writing of decimal numbers.)

$$54_8 = 4 + 5 \times 8 = 44$$

$$88_9 = 8 + 8 \times 9 = 80$$

$$A1_{16} = 1 + 10 \times 16 = 161$$

$$BBA_{13} = 10 + 11 \times 13 + 11 \times 13^2 = 2012$$

$$101_3 = 1 + 1 \times 3^2 = 10$$

$$214_6 = 4 + 1 \times 6 + 2 \times 6^2 = 82$$

$$337_8 = 7 + 3 \times 8 + 3 \times 8^2 = 223$$

Exercise 5

For which base b is the following equality true?

$$3 \times 4 = 15$$

Solution 5

The equality is true in base 7 as:

$$15_7 = 5 + 7 = 12_{10}$$

Exercise 6

Write the following numbers as sums of powers of 2:

$$45, \ 52, \ 63, \ 71, \ 89, \ 115, \ 296, \ 3012$$

Solution 6

Any positive integer can be written as a sum of powers of 2 because it has a base 2 representation. The digits in this base are 0 and 1 - their behavior is to turn "on" or "off" some power of 2. This does not work in other bases.

$$45 \ = \ 101101_2 = 2^5 + 2^3 + 2^2 + 2^0$$

$$52 \ = \ 110100_2 = 2^5 + 2^4 + 2^2$$

$$63 \ = \ 111111_2 = 2^5 + 2^4 + 2^3 + 2^2 + 2^1 + 2^0$$

$$71 \ = \ 1000111_2 = 2^6 + 2^2 + 2^1 + 2^0$$

$$89 \ = \ 1011001_2 = 2^6 + 2^4 + 2^3 + 2^0$$

$$115 \ = \ 1110011_2 = 2^6 + 2^5 + 2^4 + 2^1 + 2^0$$

$$296 \ = \ 100101000_2 = 2^8 + 2^6 + 2^4$$

$$3012 \ = \ 101111000100_2 = 2^{11} + 2^9 + 2^8 + 2^7 + 2^6 + 2^2$$

Exercise 7

Write the following numbers as algebraic sums of powers of 3 (sums where terms can be both negative and positive):

$$45, \ 52, \ 63, \ 71, \ 89, \ 115, \ 296,$$

Solution 7

Use any of the two strategies explained in Chapter 8. We will solve each example using either one or the other strategy. We will denote the representation in base 3 using negative digits (-1, 0, and 1) by $n3$

$$
\begin{aligned}
45 &= 1200_3 = 3^3 + 2 \times 3^2 = 3^3 + (3-1)3^2 \\[2mm]
&= 2 \times 3^3 - 3^2 = 3^4 - 3^3 - 3^2 \\[2mm]
52 &= 1221_3 = 1(-1)0(-1)1_{n3} = 3^4 - 3^3 - 3^1 + 3^0 \\[2mm]
63 &= 2100_3 = 1(-1)100_{n3} = 3^4 - 3^3 + 3^2 \\[2mm]
71 &= 2122_3 = 10(-1)0(-1)_{n3} = 3^4 - 3^2 - 3^0 \\[2mm]
89 &= 10022_3 = 3^4 + 2 \times 3 + 2 \\[2mm]
&= 3^4 + (3-1) \times 3 + (3-1) \\[2mm]
&= 3^4 + 3^2 - 3 + 3 - 1 = 3^4 + 3^2 - 1 \\[2mm]
115 &= 111(-1)1_{n3} = 3^4 + 3^3 + 3^2 - 3 + 3^0 \\[2mm]
296 &= 11(-1)00(-1)_{n3} = 3^5 + 3^4 - 3^3 - 3^0
\end{aligned}
$$

Exercise 8

$0.\overline{3}$ is a number in base 7. Its representation in base 10 is:

(A) 0.5

(B) $0.\overline{6}$

(C) $0.\overline{428571}$

(D) $0.\overline{3}$

Solution 8

Convert the repeating decimal into a fraction. Multiply the number by 7 to move the decimal point one place to the left:

$$7 \times 0.\overline{3} = 3.\overline{3}$$

and subtract $0.\overline{3}$ from both sides:

$$6 \times 0.\overline{3} \;=\; 3$$

$$0.\overline{3} \;=\; \frac{3}{6}$$

The correct answer choice is (A).

Exercise 9

What is the result of the operation $100000 - 1$:

1. in base 3?
2. in base 5?
2. in base 11?

Solution 9

Notice how, in any base, the result is a repdigit.

(a) 22222

(b) 44444

(c) *AAAAA*

Exercise 10

The number n^2 is written as 80 in base k. What is $8k$ in base n?

(A) 8

(B) 10

(C) 80

(D) 100

Solution 10

The correct answer choice is (D).

Write the number 80 in expanded form in base k:

$$n^2 = 8 \times k + 0$$

Since:

$$8k = n^2$$

$8k$ is written as 100 in base n.

Examples of such numbers are: $k = 18, \ 32, \ldots$.

Exercise 11

The elements of set S are the digit sums of 3-digit numbers in base 12. How many elements does the set S have?

Solution 11

The largest digit possible in base 12 is $B_{12} = 11_{10}$. A 3 digit number has a maximum digit sum of 33_{10}. Therefore, there are 34 different digit sums. However, if the number has to be a 3-digit number, then the leftmost digit cannot be zero. A digit sum of zero is not possible for a 3-digit number. Finally, there are only 33 different digit sums.

Exercise 12

Find the base x:

$$17_x + 24_x = 50_{x-1}$$

Solution 12

Write all the numbers in expanded form:

$$x + 7 + 2x + 4 = 5(x - 1) + 0$$

and solve for x:

$$
\begin{aligned}
3x + 11 &= 5x - 5 \\
16 &= 2x \\
x &= 8
\end{aligned}
$$

Exercise 13

Which of the following binary numbers is not divisible by 4?

(A) 1010101010100

(B) 1000100011100

(C) 1110011001000

(D) 1010100001010

Solution 13

Let us show that a binary number is divisible by 4 if and only if its last two digits are zero. Write a binary repunit in expanded form:

$$\underbrace{1111\dots1}_{k \text{ digits}} = 1 + 2 + 2^2 + 2^3 + \cdots + 2^k$$

If the zero-th and the first powers of 2 are "switched off" the remaining expansion admits a factor of $2^2 = 4$. This remains true if any of the other digits in higher places are "switched off."

Among the answer choices, only (D) has non-zero digits in the last two places. It is the correct answer choice.

Exercise 14

Let us extend the concept of base to numbers with decimals. An example of the expanded form of a decimal number in base 10 is:

$$231.14 = 2 \times 10^2 + 3 \times 10 + 1 + 1 \times \frac{1}{10} + 4 \times \frac{4}{10^2}$$

Convert the number 2.6 to base 5.

Solution 14

Use the expanded form:

$$2.6 = 2 + \frac{6}{10} = 2 + \frac{3}{5} = 2.3_5$$

SOLUTIONS TO PRACTICE FIVE

Do not use a calculator for any of the problems!

Exercise 1

Find the greatest common factor (greatest common divisor) of each of the following sets of numbers or expressions:

Solution 1

1. gcf(1008, 234, 1116)

$$= \text{gcf}(2^4 \times 3^2 \times 7, \ 2 \times 3^2 \times 13, \ 3^2 \times 2^2 \times 31)$$

$$= 2 \times 3^2 = 18$$

2. $\text{gcf}(2^5 \times 37^2 \times 5, \ 2^3 \times 5^2 \times 7) = 2^3 \times 5 = 40$

3. $\text{gcf}(a^2 b^3 c^4, \ a^3 b^2 c^3, \ a^4 b^3 c^2) = a^2 b^2 c^2$

4. $\text{gcf}(2^6 \times s^3, \ 2^8 \times s^2) = 2^6 \times s^2$

Exercise 2

Find the least common multiple of each of the following sets of numbers or expressions:

Solution 2

1. $\text{lcm}(155, \ 713) = \text{lcm}(23 \times 31, \ 5 \times 31) = 5 \times 23 \times 31 = 3565$

2. $\text{lcm}(1000, 1400) = \text{lcm}(2^3 \times 5^3, 2^3 \times 5^2 \times 7) = 2^3 \times 5^3 \times 7 = 7000$

3. $\operatorname{lcm}(a^2b^3c^4,\ a^3b^2c^3,\ a^4b^3c^2) = a^4b^3c^4$

4. $\operatorname{lcm}(385,\ 98,\ 55) = \operatorname{lcm}(5 \times 7 \times 11,\ 2 \times 7^2,\ 5 \times 11) = 2 \times 5 \times 7^2 \times 11 = 5390$

Exercise 3

When divided by 4, by 5, or by 6, the number N produces the remainder 3. What is the smallest possible value of N?

Solution 3

Write the integer division for each divisor:

$$
\begin{aligned}
N &= 4 \times p + 3 \\
N &= 5 \times q + 3 \\
N &= 6 \times m + 3
\end{aligned}
$$

where p, q, and m are unknown quotients.

Notice that, since the remainder is the same, it makes sense to subtract 3 from both sides of each equation:

$$
\begin{aligned}
N - 3 &= 4 \times p \\
N - 3 &= 5 \times q \\
N - 3 &= 6 \times m
\end{aligned}
$$

and find out that $N - 3$ must be divisible by 4, 5, and 6. The smallest number divisible by 4, 5, and 6 is their least common multiple:

$$
\operatorname{lcm}(4, 5, 6) = \operatorname{lcm}(2^2, 5, 2 \times 3) = 2^2 \times 3 \times 5 = 60
$$

The answer is $N = 60 + 3 = 63$.

Exercise 4

A positive integer M produces a remainder of 3 when divided by 7, a remainder of 4 when divided by 8, and a remainder of 5 when divided by 9. How many such numbers are there from 1 to 1000, inclusive?

Solution 4

We can write the integer division theorem for all three situations:

$$
\begin{aligned}
M &= 7k + 3 \\
M &= 8m + 4 \\
M &= 9p + 5
\end{aligned}
$$

Since $7, 8$, and 9 each differ from the respective remainders by 4, an idea would be to add 4 to M:

$$
\begin{aligned}
M + 4 &= 7(k + 1) \\
M + 4 &= 8(m + 1) \\
M + 4 &= 9(p + 1)
\end{aligned}
$$

Therefore, $M + 4$ is a multiple of $7, 8$ and 9. Since $7, 8$ and 9 are all *coprime*, $M + 4$ must be a multiple of $7 \cdot 8 \cdot 9 = 504$, and thus $M = 500$. There are two numbers with this property in the interval: 500 and 1000.

The correct answer is (C).

Exercise 5

The product of the lcm and the gcf of two positive integers is 391. What is the positive difference of the integers?

Solution 5

Factor 391 into primes:

$$391 = 400 - 9 = 20^2 - 3^2 = (20 + 3)(20 - 3) = 23 \times 17$$

If the two numbers are n and m, the only solution with $m > 1$ and

$n > 1$ is:

$$m = 17, \ n = 23$$

$$\gcf(17, \ 23) = 1$$

$$\text{lcm}(17, \ 23) = 391$$

$$23 - 17 = 6$$

Exercise 6

The positive integers N and M have $\text{lcm}(N, \ M) = 5670$ and $\gcf(N, \ M) = 18$. What is the smallest possible value of $N + M$?

Solution 6

Factor into primes:

$$\text{lcm}(N, \ M) = 2 \cdot 3^4 \cdot 5 \cdot 7$$

$$\gcf(N, \ M) = 2 \cdot 3^2$$

Possible solutions are:

$$N = 2 \cdot 3^2 = 18, \quad M = 2 \cdot 3^4 \cdot 5 \cdot 7 = 5670$$

$$N = 2 \cdot 3^2 \cdot 5 = 90, \quad M = 2 \cdot 3^4 \cdot 7 = 1134$$

$$N = 2 \cdot 3^2 \cdot 5 \cdot 7 = 630, \quad M = 2 \cdot 3^4 = 162$$

$$N = 2 \cdot 3^2 \cdot 7 = 126, \quad M = 2 \cdot 3^4 \cdot 5 = 810$$

The smallest possible sum is $N + M = 792$.

Exercise 7

Find the smallest sum $m + n + p$ such that:

$$\gcf(m, \ n) = 12$$

$$\gcf(m, \ p) = 30$$

$$\gcf(p, \ n) = 18$$

Solution 7

Factor into primes:

$$\gcf(m,\ n) = 12 = 2^2 \cdot 3$$
$$\gcf(m,\ p) = 30 = 2 \cdot 3 \cdot 5$$
$$\gcf(p,\ n) = 18 = 2 \cdot 3^2$$

All three numbers are multiples of $2 \cdot 3 = 6$. m and n have an additional factor of 2, m and p have an additional factor of 5, and p, n have an additional factor of 3.

The smallest numbers that satisfy the requirements are:

$$m = 2^2 \cdot 3 \cdot 5 = 60$$
$$n = 2^2 \cdot 3^2 = 36$$
$$p = 2 \cdot 3^2 \cdot 5 = 90$$

and their sum is $m + n + p = 60 + 36 + 90 = 186$.

Exercise 8

Find the smallest number that, when divided by 14 produces a remainder of 9, when divided by 12 produces a remainder of 7, and when divided by 21 produces a remainder of 16.

Solution 8

Write the integer division theorem in all cases. Denote the number by N and the unknown quotients by p, q, and m:

$$N = 14 \times p + 9$$
$$N = 12 \times q + 7$$
$$N = 21 \times m + 16$$

Notice how the differences: $14 - 9 = 5$, $12 - 7 = 5$, and $21 - 16 = 5$ are the same. This observation leads to the idea of adding 5 to N:

$$N + 5 = 14 \times p + 14$$
$$N + 5 = 12 \times q + 12$$
$$N + 5 = 21 \times m + 21$$

It is now possible to factor the right hand sides:

$$N + 5 = 14 \times (p + 1)$$
$$N + 5 = 12 \times (q + 1)$$
$$N + 5 = 21 \times (m + 1)$$

Therefore, $N + 5$ is a multiple of 14, 12, and 21. The smallest such multiple is the LCM of the numbers:

$$N + 5 = \text{LCM}(12, 14, 21)$$
$$N + 5 = \text{LCM}(2^2 \times 3, 2 \times 7, 3 \times 7)$$
$$N + 5 = 2^2 \times 3 \times 7$$
$$N + 5 = 84$$
$$N = 79$$

Indeed:

$$79 = 14 \times 5 + 9$$
$$79 = 12 \times 6 + 7$$
$$79 = 21 \times 3 + 16$$

Exercise 9

If m and n are two positive integers that satisfy $m^2 + n^2 = 8281$ and $\gcf(m,\, n) = 7$, how many different pairs $(m,\, n)$ are there?

Solution 9

The numbers must factor like this:

$$\begin{aligned} m &= 7 \cdot a \\ n &= 7 \cdot b \end{aligned}$$

The sum of the squares can be re-written as:

$$\begin{aligned} 8281 &= (7a)^2 + (7b)^2 = 49a^2 + 49b^2 = 49(a^2 + b^2) \\ 169 &= a^2 + b^2 \end{aligned}$$

The only positive integers a and b that have 169 as the sum of their squares are:

$$12^2 + 5^2 = 13^2$$

We only need to check 12, 11, and 10 because $9^2 \cdot 2 < 169$. Therefore, the only pair of positive integers that satisfies is:

$$\begin{aligned} m &= 7 \times 12 = 84 \\ n &= 7 \times 5 = 35 \end{aligned}$$

There is one single solution.

Exercise 10

If $\operatorname{lcm}(m\,n) + \gcf(m,\, n) = 113$, where m and n are two positive integers larger than 1, find $m + n$.

Solution 10

If the gcf has some prime factors, the same prime factors will divide the lcm as well. Therefore, if the gcf would be even, the lcm would also be even and the sum would be even. Since 113 is odd, it follows that the gcf must be odd and the lcm must be even.

The gcf divides the lcm, therefore we can say that:

$$\text{lcm}(m, \ n) = 2 \cdot \text{gcf}(m, \ n) \cdot k$$

The sum becomes:

$$2 \cdot k \cdot \text{gcf}(m\,n) + \text{gcf}(m, \ n) \ = \ 113$$

$$\text{gcf}(m, \ n) \times (2k + 1) \ = \ 113$$

Since 113 is prime, The gcf must be equal to 1 and the lcm must be equal to 112.

Since $\text{gcf}(m, \ n) = 1$, m and n must be coprime. Therefore, all the factors of 2 in the lcm must belong to only one of the numbers. The factorization of the lcm is: $2^4 \cdot 7$. The only solution with m and n different from 1 is:

$$m \ = \ 2^4$$

$$n \ = \ 7$$

And their sum is:

$$m + n = 16 + 7 = 23$$

Exercise 11

When dividing a positive integer N by 12, we get a remainder of 9 and when dividing the same integer by 13, we get a remainder of 5. What is the remainder when dividing N by 12×13?

Solution 11

Write the integer divisions by 12 and by 13:

$$N \ = \ 12 \times k + 9$$

$$N \ = \ 13 \times j + 5$$

Multiply the first equality by 13 and the second one by 12:

$$13 \cdot N \ = \ 12 \cdot 13 \cdot k + 117$$

$$12 \cdot N \ = \ 13 \cdot 12 \cdot j + 60$$

and subtract these equations:

$$N = 12 \cdot 13(k + j) + 117 - 60$$
$$N = 12 \cdot 13(k + j) + 57$$

Since $12 \times 13 = 156$ and $57 < 156$, we conclude that the required remainder is 57.

Exercise 12

How many positive integers smaller than 5000 produce a remainder of 7 when divided by 18, 21, and 30?

(A) 0

(B) 2

(C) 4

(D) 7

Solution 12

The correct answer is (D).

Denote the number with this property by N. Then:

$$N = 18 \times p + 7$$
$$N = 21 \times q + 7$$
$$N = 30 \times m + 7$$

Subtract 7 from N:

$$N - 7 = 18 \times p$$
$$N - 7 = 21 \times q$$
$$N - 7 = 30 \times m$$

Therefore, $N - 7$ is a multiple of 18, 21, and 30. The smallest such multiple is:

$$N - 7 = \text{LCM}(18, 21, 30) = \text{LCM}(2 \times 3^2, 3 \times 7, 2 \times 3 \times 5) = 2 \times 3^2 \times 7 \times 5 = 630$$

And the smallest possible value of N is 637. There are 7 values for N. The largest is:

$$N = 7 \cdot 630 + 7 = 4417$$

Exercise 13

If a and b are positive integers with $\text{lcm}(a, b) = 840$ and $\text{gcf}(a, b) = 28$, find $a \times b$.

Solution 13

Use the property explained in the lecture notes:

$$\text{lcm}(a, b) \times \text{gcf}(a, b) = a \times b$$
$$840 \times 28 = 23,520$$

Exercise 14

If $\text{lcm}(a, b) = 2^2 \cdot 3^2$ and $\text{gcf}(a, b) = 3$, which of the following must be true:

(A) a and b are coprime.

(B) a and b are both even.

(C) a and b are both odd.

(D) a and b have different parity.

Solution 14

(A) If two numbers a and b are coprime, then $\text{gcf}(a, b) = 1$. (A) is false.

(B) If both numbers a and b are even, then there is a factor of at least 2 in the gcf. Since the gcf is odd, (B) is false.

(C) If both numbers a and b are odd, then there is no factor of 2 in the lcm. Since the lcm is even, (C) is false.

(D) Since the gcf is odd, both factors of 2 must belong to only one of the numbers. One of the numbers is odd and the other is even. (D) is true.

Solutions to Practice Six

Do not use a calculator for any of the problems!

Exercise 1

If m and k are non-zero integers, how many solutions does the equation have:

$$km + m = 6$$

(A) 1

(B) 2

(C) 3

(D) 7

Solution 1

The correct answer is (D). Factor the left side:

$$m(k + 1) = 2 \times 3$$

and list all the possible matches:

Case 1	$m = 3$ $k + 1 = 2$	$m = 3$ $k = 1$	valid solution
Case 2	$m = 2$ $k + 1 = 3$	$m = 2$ $k = 2$	valid solution
Case 3	$m = -3$ $k + 1 = -2$	$m = -3$ $k = -3$	valid solution
Case 4	$m = -2$ $k + 1 = -3$	$m = -2$ $k = -4$	valid solution
Case 5	$m = 1$ $k + 1 = 6$	$m = 1$ $k = 5$	valid solution
Case 6	$m = -1$ $k + 1 = -6$	$m = -1$ $k = -7$	valid solution
Case 7	$m = 6$ $k + 1 = 1$	$m = 6$ $k = 0$	invalid solution
Case 8	$m = -6$ $k + 1 = -1$	$m = -6$ $k = -2$	valid solution

Exercise 2

If m and k are integers, what is the largest possible value of $m \cdot k$:

$$km - m + k = 16$$

(A) 12

(B) 16

(C) 28

(D) 32

Solution 2

The correct answer is (C). Force a factoring on the left side as follows:

$$
\begin{aligned}
m(k - 1) + k - 1 &= 16 - 1 \\
(k - 1)(m + 1) &= 15 \\
(k - 1)(m + 1) &= 3 \times 5
\end{aligned}
$$

and list all possible matches:

Case 1	$m + 1 = 5$ $k - 1 = 3$	$m = 4$ $k = 4$	$m \cdot k = 16$
Case 2	$m + 1 = 3$ $k - 1 = 5$	$m = 2$ $k = 6$	$m \cdot k = 12$
Case 3	$m + 1 = 1$ $k - 1 = 15$	$m = 0$ $k = 16$	$m \cdot k = 0$
Case 4	$m + 1 = 15$ $k - 1 = 1$	$m = 14$ $k = 2$	$m \cdot k = 28$
Case 5	$m + 1 = -5$ $k - 1 = -3$	$m = -6$ $k = -2$	$m \cdot k = 12$
Case 6	$m + 1 = -3$ $k - 1 = -5$	$m = -4$ $k = -4$	$m \cdot k = 16$
Case 7	$m + 1 = -1$ $k - 1 = -15$	$m = -2$ $k = -14$	$m \cdot k = 28$
Case 8	$m + 1 = -15$ $k - 1 = -1$	$m = -16$ $k = 0$	$m \cdot k = 0$

Exercise 3

What is the largest number of positive integers larger than 1 that have a product of 5400?

Solution 3

Factor 5400 into primes:

$$5400 = 2^3 \cdot 3^3 \cdot 5^2$$

The largest number of factors is given by the number of prime factors, as prime factors cannot be factored into even smaller factors. The largest number of numbers that have a product of 5400 is $3+3+2 = 8$.

Exercise 4

Each side of a convex quadrilateral is labeled with a positive number. Each vertex of the quadrilateral is labeled with a number that is the product of the numbers on the two sides that meet at the vertex. What is the sum of the numbers that label the sides?

(A) 4

(B) 7

(C) 9

(D) 15

Solution 4

Denote the numbers that label the sides by a, b, c, and d:

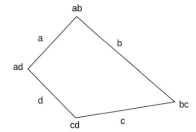

The numbers that label the vertices are the positive integers ab, bc, cd, and ad. Their sum is:

$$ab + bc + cd + ad = 14$$

Factor the left and right sides:

$$b(a + c) + d(a + c) = 14$$

$$(a + c)(b + d) = 2 \times 7$$

The possibilites for the sum $a + b + c + d$ are:

Case 1	$a + c = 7$ $b + d = 2$	$a + b + c + d = 9$
Case 2	$a + c = 14$ $b + d = 1$	$a + b + c + d = 15$

Case 2 is invalid, since $b + d = 1$ cannot be satisfied by two positive

122

values for b and d. The correct answer choice is (C).

Exercise 5

How many pairs of different positive integers have a sum of 67?

(A) 33

(B) 34

(C) 66

(D) 67

Solution 5

The correct answer choice is (A).

Denote the integers by a and b. Then:

$$a + b = 67$$

iwth $a > 0, b > 0$, and $a \neq b$.

$$1 + 66 = 67$$
$$\cdots = \cdots$$
$$33 + 34 = 67$$

The remaining pairs are the same only in reverse order. Since addition is commutative, only 33 pairs of terms are distinct.

Exercise 6

The set M contains a number of distinct (i.e. different) odd integers. We form all possible positive differences among their squares. At least how many integers should there be in the set M so as to have at least two differences that end in the same digit?

(A) 3

(B) 4

(C) 5

(D) there is not enough information

Solution 6

Odd integers end in 1, 3, 5, 7, or 9. Their squares end in 1, 9, 5, 9, or 1. Therefore, there are only 3 distinct choices for the last digit of their squares. According to the Pigeonhole principle, there have to be at least 4 numbers in set M. The correct answer choice is (B).

Exercise 7

The set J contains fourth powers of N distinct integers. How many possible values are there for the last digit of the product of the numbers in J?

(A) 4

(B) 6

(C) 8

(D) 9

Solution 7

Let us make a table with the last digit of the integer powers of consecutive numbers:

last digit of N	0	1	2	3	4	5	6	7	8	9
last digit of N^2	0	1	4	9	6	5	6	9	4	1
last digit of N^4	0	1	6	1	6	5	6	1	6	1

If all the elements of J are fourth powers, then there are only 4 possible values for the last digit of their product: 0, 1, 5, and 6.

Exercise 8

The middle number among three consecutive integers is a perfect square. From the choices below, select the gcf of the products of all such sets of three numbers.

(A) 1

(B) 4

(C) 12

(D) 60

Solution 8

Three consecutive numbers can be either (odd, even, odd) or (even, odd, even). If the middle number is an even perfect square, then it must be divisible by 4. If the middle number is an odd perfect square, then the other two numbers are both even and the product is divisible by 4. Therefore, either way the product of the three numbers is divisible by 4.

Since among three consecutive numbers one is surely a multiple of 3, the product is also a multiple of 3.

Looking at the table with last digits of perfect squares:

last digit of N	0	1	2	3	4	5	6	7	8	9
last digit of N^2	0	1	4	9	6	5	6	9	4	1

we see that a perfect square is either divisible by 5 or one has to subtract 1 or add 1 to it to make it divisible by 5. This means that one of: $N^2 - 1$, N^2, or $N^2 + 1$ is a multiple of 5. Therefore, their product is a multiple of 5.

The correct answer choice is (D).

Exercise 9

Write 6^{2015} as a sum of three perfect cubes.

Solution 9

Split advantageously the factors of 6:

$$
\begin{aligned}
6^{2015} &= 6^2 \cdot 6^{2013} \\[2mm]
&= 36 \cdot 6^{671 \cdot 3} \\[2mm]
&= (1 + 8 + 27) \cdot \left(6^{671}\right)^3 \\[2mm]
&= \left(1^3 + 2^3 + 3^3\right) \cdot \left(6^{671}\right)^3 \\[2mm]
&= \left(6^{671}\right)^3 + 2^3 \cdot \left(6^{671}\right)^3 + 3^3 \cdot \left(6^{671}\right)^3 \\[2mm]
&= \left(6^{671}\right)^3 + \left(2 \cdot 6^{671}\right)^3 + \left(3 \cdot 6^{671}\right)^3
\end{aligned}
$$

Exercise 10

In a $N \times M$ grid of white square grid cells, three adjacent cells that are either in the same row or in the same column are colored blue. If there are 416 ways to do this, how many grid cells does the grid have?

(**A**) 208

(**B**) 240

(**C**) 416

(**D**) 832

Solution 10

Three adjacent and collinear squares can fit on an $N \times M$ grid in $M(N-2) + N(M-2)$ ways:

126

2 possible colorings

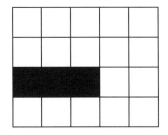
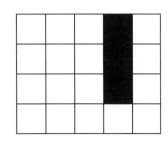

Apply the fundamental theorem of arithmetic to solve this Diophantine equation:

$$M(N-2) + N(M-2) = 416$$

Manipulate the expression and factor it out:

$$
\begin{aligned}
MN - 2M + NM - 2N &= 416 \\
2MN - 2M - 2N &= 416 \\
MN - M - N &= 208 \\
MN - M - N + 1 &= 209 \\
M(N-1) - (N-1) &= 209 \\
(M-1)(N-1) &= 209 \\
(M-1)(N-1) &= 11 \cdot 19
\end{aligned}
$$

The grid must have dimensions 12 and 20 and a total of 240 grid cells.

The correct answer is (B).

Exercise 11

Dina and Lila are playing cards. Dina has N cards. Lila has either 2 cards more or 2 cards less than Dina. If Dina does not know how many cards Lila is holding, then N is:

(A) 3 or more

(B) 3 or less

(C) a multiple of 4

(D) a multiple of $N + 2$

Solution 11

If Dina had 1 card, she would know for sure that Lila has 3 cards. If Dina had 2 cards, she would know for sure that Lila has 4 cards. For any other numbers of cards that Dina could have, there are two possible choices for the number of Lila's cards.

The correct answer is (A).

Exercise 12

What are the values of the integers k and p?

$$2^k + 3^p = 857$$

Solution 12

How can the sum end in 7?

Powers of 2 end in: 2, 4, 8, 6, while powers of 3 end in: 3, 9, 7, 1. For the sum to end in 7 we must have:

- a power of 2 that ends in 6 and a power of 3 that ends in 1;
- a power of 2 that ends in 4 and a power of 3 that ends in 3;
- a power of 2 that ends in 8 and a power of 3 that ends in 9.

Since $2^{10} > 857$ and $3^7 > 857$, we have $k \leq 9$ and $p \leq 7$. The plausible pairs form the three following sequences:

$$(2^4,\ 3^4),\ (2^8,\ 3^4)$$

$$(2^2,\ 3^1),\ (2^6,\ 3^5),\ (2^2,\ 3^5),\ (2^6,\ 3^5)$$

$$(2^3,\ 3^2),\ (2^7,\ 3^2),\ (2^3,\ 3^6),\ (2^7,\ 3^6)$$

Since the sum is not a very large number, we should be able to find the required sum among the first few terms. Indeed, we find that:

$$2^7 + 3^6 = 857$$

$k = 7$ and $p = 6$.

Exercise 13

In a storm relief medical facility, a certain amount of supplies is needed on average for each case. If there were 3 cases fewer per day, the 77 existing supply packs would last 5 days longer. How many cases are seen each day? If there are several solutions, list them all.

Solution 13

An equation is written as follows:

$$(D + 5)(P - 3) = 77$$

where D is the number of days and P is the number of cases examined each day. This equation is easily recognized as Diophantine since the number of persons and the number of days must be positive integers. This equation can be solved by factoring as follows:

$$(D + 5)(P - 3) = 7 \cdot 11$$

List only the solutions that lead to positive values for D and P:

Case 1	$D + 5 = 7$	$D = 2$
	$P - 3 = 11$	$P = 14$
Case 2	$D + 5 = 11$	$D = 6$
	$P - 3 = 7$	$P = 10$
Case 3	$D + 5 = 77$	$D = 72$
	$P - 3 = 1$	$P = 4$

129

Exercise 14

If x and y have integer values, how many distinct solutions does the equation have?

$$|x - y| + y^2 = 9$$

Solution 14

Since perfect squares and absolute values are both positive, there are only the following cases:

i. $0 + 9 = 9$. Therefore, $y^2 = 9$ and $y = 3$ or $y = -3$. If $y = 3$, then $x = 3$. If $y = -3$, then $x = -3$.

ii. $5 + 4 = 9$. Therefore, $y^2 = 4$ and $y = 2$ or $y = -2$. If $y = 2$, then $|x - 2| = 5$ and $x = 7$, $x = -3$ are both solutions. If $y = -2$ then $|x + 2| = 5$ and $x = 3$, $x = -7$ are both solutions.

ii. $8 + 1 = 9$. Therefore, $y^2 = 1$ and $y = 1$ or $y = -1$. If $y = 1$, then $|x - 1| = 8$, and $x = 9$, $x = -7$ are both solutions. If $y = -1$, then $|x + 1| = 8$, and $x = 7$, $x = -9$ are both solutions.

iv. $9 + 0 = 9$. Therefore $y^2 = 0$ and $y = 0$. Then, $|x| = 9$ so $x = 9$, $x = -9$ are both solutions.

There are 12 distinct solutions.

Solutions to Miscellaneous Practice

Do not use a calculator for any of the problems!

Exercise 1

For how many integer values of k is the fraction an integer:

$$\frac{10 + k}{k - 1}$$

(A) 0

(B) 1

(C) 2

(D) 4

Solution 1

Apply the division algorithm to separate the integers from the proper fraction:

$$\frac{10 + k}{k - 1} = \frac{11 + k - 1}{k - 1}$$

$$= \frac{11}{k - 1} + \frac{k - 1}{k - 1}$$

$$= \frac{11}{k - 1} + 1$$

The problem is now reduced to finding values for $k - 1$ that are divisors of 11:

$$k - 1 \in \{-11, -1, 1, 11\}$$

The following values of k fulfill the requirement:

$$k \in \{-10, 0, 2, 12\}$$

There are 4 possible values for k. The correct answer is (D).

Exercise 2

When dividing a positive integer by 17, we obtain a quotient equal to the remainder. What is the largest digit sum the number can have?

(A) it can have any digit sum

(B) 17

(C) 18

(D) 27

Solution 2

Denote the number by N and write the integer division:

$$N = 17 \times R + R$$

Factor out R:

$$N = R \times (17 + 1) = 18 \times R$$

Since R is a remainder, it can only have integer values from 0 to 16. The largest digit sums are obtained for:

$$18 \times 11 = 198 \qquad S(198) = 18$$
$$18 \times 16 = 288 \qquad S(288) = 18$$

The correct answer is (C).

Exercise 3

The number N is formed by writing the numbers from 1 to 100 in increasing order in a row:

$$N = 12345 \cdots 9899100$$

Erase 16 digits so as to obtain the smallest possible number. What is the sum of the digits you erased?

Solution 3

To make the result as small as possible, one has to have as small digits as possible in the places with larger place value. However, if we make 0 the first digit, we manage to decrease the number of digits without actually erasing an extra digit. Therefore, we first erase the digits that separate the leftmost digit from the next digit of 0:

$$\cancel{123456789}101112\cdots 100$$

The sum of the digits we erased is:

$$1 + 2 + 3 + 4 + \cdots + 9 + 1 = \frac{9 \cdot 10}{2} + 1 = 46$$

With this, we have erased 10 digits and we now have a digit of zero in the leftmost position. To make the second digit from the left also zero we would have to erase more than 16 digits in total. We have permission to erase only 16 digits of which we have already erased 11 at the previous step. Therefore, our strategy is to obtain as many digits of 1 in the leftmost positions. We have 5 digits to erase:

$$1011\cancel{121314151}617181920\cdots 99100$$

The sum of the digits we erased at the second step is:

$$2 + 3 + 4 + 5 + 6 = 20$$

The sum of all the 16 digits we erased is: $46 + 20 = 66$

Exercise 4

If the sum of 45 positive consecutive integers is a perfect square, what is the smallest possible value of the median number?

(A) 5

(B) 20

(C) 25

(D) 45

Solution 4

Write the consecutive integers in the form:

$$k - 22, \ k - 21, \ k - 20, \ \cdots, k, \cdots, k + 20, \ k + 21, \ k + 22$$

The median integer is k and the sum of the integers is $45 \times k$.

The smallest k that makes $45 \times k$ a perfect square is 5 but this value does not lead to a set of positive numbers. Indeed, the first number would be $5 - 22 = -17$.

The next larger value for k is 20 for which the sum becomes equal to $3^2 \times 2^2 \times 5^2 = 30^2 = 900$. In this case, as well, the first number would be negative: -2.

The next larger value for k is 45. For this value, all the terms are positive. The correct answer is (D).

Exercise 5

Order the following numbers by increasing value:

$$0.\overline{9374}, \ 0.9\overline{374}, \ 0.93\overline{74}, \ 0.937\overline{4}, \ 0.93\overline{7}, \ 0.9\overline{43}$$

Solution 5

In increasing order, the numbers are:

$$0.93\overline{7}, \ 0.9\overline{374}, \ 0.937\overline{4}, \ 0.93\overline{74}, \ 0.\overline{9374}, \ 0.9\overline{43}$$

Exercise 6

How many pairs of positive integers have a sum of 305 and a remainder of 4 when the larger number is divided by the smaller one?

Solution 6

Assume the numbers are a and b with $a > b$. Then, the integer division is:

$$a = b \times q + 4$$

where q is some quotient and the remainder is, as required, 4. The sum of the numbers is:

$$
\begin{aligned}
b \times q + 4 + b &= 305 \\
b \times q + b &= 301 \\
b \times q + b &= 7 \times 43 \\
b \times (q + 1) &= 7 \times 43
\end{aligned}
$$

Because the prime factorization is unique, b can be one of: 1, 7, 43, 301. The values of a corresponding to these values for b are: 304, 298, 262, 4. Of these values, only the first three are larger than the corresponding value for b, forming the pairs: $(304, 1), (298, 7)$ and $(262, 43)$. However, b must be larger than the remainder. Therefore, the solution is formed only by the pairs: $(298, 7)$ and $(262, 43)$. The set of solutions has 2 elements.

Exercise 7

Write 5^{2013} as a sum of two perfect squares.

Solution 7

Split advantageously the factors of 5:

$$
\begin{aligned}
5^{2013} &= 5^3 \cdot 5^{2010} \\[2mm]
&= 125 \cdot 5^{1005 \cdot 2} \\[2mm]
&= (4 + 121) \cdot \left(5^{1005}\right)^2 \\[2mm]
&= (2^2 + 11^2) \cdot \left(5^{1005}\right)^2 \\[2mm]
&= 2^2 \cdot \left(5^{1005}\right)^2 + 11^2 \cdot \left(5^{1005}\right)^2 \\[2mm]
&= \left(2 \cdot 5^{1005}\right)^2 + \left(11 \cdot 5^{1005}\right)^2
\end{aligned}
$$

Exercise 8

30 consecutive numbers have a positive sum smaller than 465. What is the product of the 30 consecutive numbers?

Solution 8

The product is zero. If the sequence of 30 consecutive numbers starts with the number 1, their sum is:

$$\frac{30 \cdot 31}{2} = 465$$

Since the sum is smaller than 465, the sequence must start with a number smaller than 1. Since the sum is positive, the sequence must have some positive and some negative terms. Therefore, it has a term equal to 0. This term makes the product zero.

Exercise 9

30 consecutive numbers have a positive sum smaller than 465. What is the largest such sum?

Solution 9

The largest sum is obtained by adding the consecutive numbers:

$$0 + 1 + 2 + 3 + \cdots + 29 = \frac{29 \cdot 30}{2} = 435$$

Exercise 10

If P and Q are positive integers and:

$$(Q + 5) \cdot (P - 7) = 2 \cdot Q \cdot P$$

How many possible solutions are there?

Solution 10

We have to factor the equation so that the unknown numbers P and Q are both on the same side of the equation:

$$
\begin{aligned}
Q \cdot P + 5 \cdot P - 7 \cdot Q - 35 &= 2 \cdot P \cdot Q \\
5 \cdot P - 7 \cdot Q - 35 &= P \cdot Q \\
5 \cdot P - 7 \cdot Q - P \cdot Q &= 35 \\
P \cdot (5 - Q) - 7 \cdot Q &= 35 \\
P \cdot (5 - Q) + 35 - 7 \cdot Q &= 70 \\
P \cdot (5 - Q) + 7 \cdot (5 - Q) &= 70 \\
(P + 7) \cdot (5 - Q) &= 2 \cdot 5 \cdot 7
\end{aligned}
$$

There are only two solutions with positive values for P and Q:

$P + 7 = 35$	$P = 28$
$5 - Q = 2$	$Q = 3$
$P + 7 = 70$	$P = 63$
$5 - Q = 1$	$Q = 4$

Exercise 11

    i. What is the sum of all 3-digit positive integers with digits 1, 2 and 3, if each digit is used once in each number?

    ii. What is the sum of all 3-digit positive integers with digits 1, 2 and 3, if digits can be used repeatedly?

Solution 11

    i. Each of the digits can occur in any of the three available place values. In each place value, each digit occurs twice. For example, the digit 1 occurs twice in the hundreds, twice in the tens and twice in the units places:

$$123$$

$$132$$

$$213$$

$$312$$

$$321$$

$$231$$

The digit 1 contributes to the sum of all these numbers:

$$1 \times 2 \times 111$$

Therefore, the sum is:

$$
\begin{aligned}
S &= 1 \times 2 \times 111 + 2 \times 2 \times 111 + 3 \times 2 \times 111 \\
&= (1 + 2 + 3) \times 2 \times 111 \\
&= 6 \times 2 \times 111 \\
&= 12 \times 111 = 2331
\end{aligned}
$$

    ii. Similarly,

$$S = (1 + 2 + 3) \cdot 9 \cdot 111 = 5994$$

Exercise 12

What is the sum of all 5-digit positive integers with digits 1, 3, 5, 7, and 9, if each digit is used once in each number?

Solution 12

Each digit occurs in each possible place value $4! = 24$ times. The contribution of each digit d to the sum is:

$$d \times 24 \times 11111$$

The sum is:

$$
\begin{aligned}
S &= (1 + 3 + 5 + 7 + 9) \times 24 \times 11111 \\
&= 25 \times 24 \times 11111 \\
&= 100 \times 6 \times 11111 \\
&= 600 \times 11111 \\
&= 6666600
\end{aligned}
$$

Exercise 13

What is the sum of all 4-digit numbers with 2 digits of 2 and 2 digits of 7?

Solution 13

Here is an example of such a number: 2277.

The digit 2 may occur in the leftmost place in 3 different ways:

$$2277, \ 2727, \ 2772$$

The same is true for the digit 7. The contribution to the sum of the possible ways to assign a digit to the leftmost place is:

$$(2 + 7) \times 3 \times 1000$$

139

The computation is similar for any place value. Therefore, the sum is:

$$
\begin{aligned}
S &= (2+7) \times 3 \times 1111 \\
&= 27 \times 1111 \\
&= 29997
\end{aligned}
$$

Exercise 14

What is the sum of all the positive integers that have a digit sum and a digit product both equal to 14?

Solution 14

Factor 14 into primes:

$$14 = 2 \times 7$$

Therefore two of the digits must be 2 and 7. Since $2 + 7 = 9$ we have to add $14 - 9 = 5$ digits of 1 to the number, to have the same digit sum as digit product:

$$2711111$$

The numbers with this property have the same digits as the example above only placed in different place values. Each of the digits can appear in each place value in six different ways:

$$2711111$$

$$2171111$$

$$2117111$$

$$2111711$$

$$2111171$$

$$2111117$$

Therefore, the sum of all possible numbers with this property is:

$$
\begin{aligned}
S &= 6 \times (2 + 7 + 1 + 1 + 1 + 1 + 1) \times 1111111 \\
&= 6 \times 14 \times 1111111 \\
&= 84 \times 1111111 \\
&= 93,333,324
\end{aligned}
$$

Exercise 15

What is the sum of the first 25 decimal digits of each of the following numbers:

i. $\sqrt{0.\underbrace{99\cdots9}_{25\text{ digits}}}$

ii. $\sqrt{0.\underbrace{11\cdots1}_{25\text{ digits}}}$

iii. $\sqrt{0.\underbrace{44\cdots4}_{25\text{ digits}}}$

Solution 15

We use the properties of repeating non-terminating fractions:

$$
0.\overline{9} = 1
$$

and we write the inequality:

$$
\underbrace{0.99\cdots9}_{25\text{ digits}} < 0.\overline{9} = 1
$$

Since the square of any positive real number smaller than 1 is smaller than the number (if $0 < q < 1$ then $0 < q^2 < q < 1$):

$$
\underbrace{0.99\cdots9}_{25\text{ digits}} < \sqrt{\underbrace{0.99\cdots9}_{25\text{ digits}}} < 0.\overline{9} = 1
$$

The number we seek has 25 digits of 9 after the decimal point followed by other digits. The sum of these 25 digits is $25 \cdot 9 = 225$.

We can apply the same process to the following numbers.

Dividing by 3, we obtain:

$$\underbrace{0.33\cdots3}_{25 \text{ digits}} < \sqrt{\frac{1}{9} \cdot \underbrace{0.99\cdots9}_{25 \text{ digits}}} < \frac{1}{3}$$

$$\underbrace{0.33\cdots3}_{25 \text{ digits}} < \sqrt{\underbrace{0.11\cdots1}_{25 \text{ digits}}} < 0.\overline{3} = \frac{1}{3}$$

In this case, the first 25 digits are all equal to 3 and their sum is 75.

Multiplying the previous inequalities by 2, we obtain:

$$\underbrace{0.66\cdots6}_{25 \text{ digits}} < \sqrt{\underbrace{0.44\cdots4}_{25 \text{ digits}}} < 0.\overline{6} = \frac{2}{3}$$

The first 25 decimal digits are all equal to 6 and their sum is 150.

Exercise 16

In base 5, how many positive integers smaller than 1000 are multiples of 4?

Solution 16

$1000_5 = 125_{10}$. There are 31 positive multiples of 4.

Exercise 17

The expression $b^2 + b + 1$ is:

(A) always even

(B) always odd

(C) sometimes even

Solution 17

Since
$$b^2 + b + 1 = b(b + 1) + 1$$

where $b(b + 1)$ is the product of two consecutive integers and is, therefore, always even. It follows that the expression $b^2 + b + 1$ is always odd, for any integer value of b.

The correct answer is (B).

Exercise 18

For how many integer values of t is the fraction below reducible?

$$\frac{t^2 - t + 4}{t^2 + t}$$

(A) 2

(B) 5

(C) 8

(D) an infinity

Solution 18

The correct answer is (D).

The denominator is an even number, since it can be factored into a product of two consecutive integers:

$$t^2 + t = t(t + 1)$$

The numerator is also an even number since it can be rewritten in the form:

$$t(t - 1) + 4$$

which is a sum of two even numbers (the product of two consecutive integers $t(t - 1)$ and 4).

Since both numerator and denominator are even, the fraction is reducible by 2 for any integer t different from zero and -1.

Exercise 19

How many of the following fractions are irreducible?

$$\frac{589}{2}, \frac{589}{3}, \cdots \frac{589}{4}, \frac{589}{589}$$

Solution 19

Factor 589 into primes: $589 = 19 \times 31$. Such a fraction will be reducible if the denominator has a factor that is also a factor of the numerator. The number of multiples of 19 that can be denominators is:

$$\frac{589}{19} = 31$$

The number of multiples of 31 is:

$$\frac{589}{31} = 19$$

The number of multiples of 19×31 is 1.

The number of denominators that have common factors with the numerator 589 is $19 + 31 - 1 = 49$. There are 49 reducible fractions. The total number of fractions is 588. Therefore, there are $588 - 49 = 539$ irreducible fractions in the set.

Exercise 20

x, y, and z are unknown digits in the base-10 number $4xy77z$. What are the digits x, y, and z that make the number divisible by 888?

Solution 20

Factor 888 into primes: $888 = 2^3 \times 3 \times 37$. For the number to be divisible by 8 the last three digits must form a number divisible by 8. Divide $77z$ by 8 to see that z must be 6.

Now, the number $4xy776$ must be divisible by 111. Let us cast out all the 111s:

$$4xy776 = 400776 + xy \times 1000$$

$$400776 = 3610 \times 111 + 66$$

$$1000 = 9 \times 111 + 1$$

Therefore:

$$4xy776 = 3610 \times 111 + 9xy \times 111 + 66 + xy$$

The number is divisible by 111 if $66 + xy$ is divisible by 111. This can only happen if $xy = 45$.

The required number is 445776.

Exercise 21

How many distinct pairs of positive numbers have an lcm of 25?

(A) 1

(B) 2

(C) 3

(D) 4

Solution 21

The pairs of positive numbers that have an lcm of 25 are:

$$(1, \ 25)$$

$$(5, \ 25)$$

$$(25, \ 25)$$

There is a total of 3 pairs. The correct answer is (C).

Exercise 22

How many distinct pairs of positive numbers have an lcm of 36?

(**A**) 9

(**B**) 12

(**C**) 13

(**D**) 25

Solution 22

Factor into primes: $36 = 2^2 \times 3^2$. We can distribute the factors of 2 and 3 among the two numbers as follows:

	A	B	
	$2^0, 3^0$	$2^2, 3^2$	(1, 36)
	$2^0, 3^1$	$2^2, 3^2$	(3, 36)
	$2^0, 3^2$	$2^2, 3^2$	(9, 36)
	$2^0, 3^2$	$2^2, 3^1$	(9, 12)
	$2^0, 3^2$	$2^2, 3^0$	(9, 4)
	$2^1, 3^0$	$2^2, 3^2$	(2, 36)
	$2^1, 3^1$	$2^2, 3^2$	(6, 36)
	$2^1, 3^2$	$2^2, 3^2$	(18, 36)
	$2^1, 3^2$	$2^2, 3^1$	(18, 12)
	$2^1, 3^2$	$2^2, 3^0$	(18, 4)
	$2^2, 3^2$	$2^2, 3^0$	(36, 4)
	$2^2, 3^2$	$2^2, 3^1$	(36, 12)
	$2^2, 3^2$	$2^2, 3^2$	(36, 36)

Exercise 23

It took Amon, the elephant, several days to eat all the melons in Sarawut's patch. If Amon had eaten 20 more melons per day during twice the number of days, he would have eaten 265 more melons. How many melons did Amon eat each day, on average?

Solution 23

Denote the number of days with D and the number of melons eaten each day with M. Then we have:

$$D \cdot M + 265 = 2 \cdot D \cdot (M + 20)$$

Simplify the equation:

$$DM + 265 = 2DM + 40D$$
$$DM + 40D = 265$$

This is a factorable Diophantine equation because the number of melons and the number of days must have positive integer values.

$$D(M + 40) = 265$$
$$D(M + 40) = 5 \cdot 53$$

Analyze the following choices that assume the variables D and M to be positive:

$M + 40 = 53$	$M = 13$
$D = 5$	$D = 5$
$M + 40 = 265$	$M = 225$
$D = 1$	$D = 1$

We retain the solution with $D = 5$ since $D = 1$ does not represent "a few" days.

Amon ate 13 melons a day for 5 days.

Exercise 24

What is the largest possible digit sum of the digit sum of a 4-digit number?

Solution 24

The digit sum of a 4-digit number ranges between 1 and $9 \times 4 = 36$. The largest digit sum among the integers between 1 and 36 is 11, the digit sum of the number 29.

Exercise 25

What is the largest possible digit product of the digit sum of a 4-digit number?

Solution 25

The digit sum of a 4-digit number ranges between 0 and $9 \times 4 = 36$. The digit products of the numbers from 0 to 36 range from 0 to 18.

Exercise 26

The following numbers are in base 6. Which one of them is divisible by 5?

(A) 1234

(B) 1235

(C) 1254

(D) 1255

Solution 26

We can prove the following rule of divisibility by 5 in base 6: a number in base 6 is divisible by 5 if the sum of its digits is divisible by 5.

First, let us sketch a proof by using a random numeric example. If $N = 5133$ in base 6, it can be written in expanded form. The exapanded form can be used to cast out the multiples of 5. Write in expanded form, replace the 6 by $5 + 1$, and separate all the multiples of 5:

$$
\begin{aligned}
4234_6 &= 4 \cdot 6^3 + 2 \cdot 6^2 + 3 \cdot 6 + 4 \\
&= 4 \cdot (5+1)^3 + 2 \cdot (5+1)^2 + 3 \cdot (5+1) + 4 \\
&= 4 \cdot (5^3 + 3 \cdot 5^2 + 3 \cdot 5 + 1) + 2 \cdot (5^2 + 2 \cdot 5 + 1) + 3 \cdot 5 + 3 + 4 \\
&= 5 \cdot (4 \cdot 5^2 + 4 \cdot 3 \cdot 5 + 4 \cdot 3 + 2 \cdot 5 + 2 \cdot 2 + 3 \cdot 5) + 4 + 2 + 3 + 4
\end{aligned}
$$

This means the number 4234_6 is the sum of a multiple of 5 and the number $4 + 2 + 3 + 4$ which is the digit sum of the number. If the digit sum is divisible by 5 so is the number.

A formal proof can be written along the same lines but would be more cumbersome because of having an arbitrary number of digits and literals for digits.

For the given numbers, the digit sums are $1+2+3+4 = 10$, $1+2+3+5 = 11$, $1 + 2 + 5 + 4 = 12$, and $1 + 2 + 5 + 5 = 13$. Of these, only the first one is divisible by 5. The correct answer is (A).

Exercise 27

If a, b, c, and m are base-10 digits, which of the following must be a common factor of the following numbers $abcabc$ and $m77m$?

(A) a

(B) $a + m$

(C) $7a$

(D) 5

(E) 77

Solution 27

Use the digit patterns to find actual numeric factors:

$$abcabc = abc \cdot 1001 = abc \cdot 7 \cdot 11 \cdot 13$$

$$m77m = m \cdot 1001 + 770 = m \cdot 7 \cdot 11 \cdot 13 + 7 \cdot 11 \cdot 10$$

Regardless of the values of the unknown digits, 77 is a factor of both integer numbers. The correct answer choice is (E).

Exercise 28

How many digits of 5 must the following number have in order to be a perfect square:

$$1 \underbrace{555 \ldots 55}_{k \text{ digits}}$$

Solution 28

0 is the only possibility.

Divide the number by 4 to find that, for any number of digits of 5, the remainder of the division is 3.

No perfect square can have a remainder of 3 when divided by 4.

Exercise 29

Three consecutive positive integers have an lcm of 976500. What is the smallest possible product of the three numbers?

Solution 29

Factor into primes:

$$976500 = 2^2 \cdot 3^2 \cdot 5^3 \cdot 7 \cdot 31$$

The three factors of 5 belong to a single number, since there is at most one multiple of 5 among three consecutive numbers. Therefore, one of the numbers is a multiple of 125. The smallest such multiple is 125 and the set of numbers could be $\{125, 126, 127\}$, $\{124, 125, 126\}$, or $\{123, 124, 125\}$. If none of these three sets has the required lcm we will go to the next multiple of 125, which is 250.

$$\text{lcm}(125,\ 126,\ 127) = \text{lcm}(5^3,\ 2 \cdot 3^2 \cdot 7,\ 127) = 2,000,250$$

$$\text{lcm}(124,\ 125,\ 126) = \text{lcm}(2^2 \cdot 31, 5^3, 2 \cdot 3^2 \cdot 7) = 976,500$$

$$\text{lcm}(123,\ 124,\ 125) = \text{lcm}(3 \cdot 41,\ 2^2 \cdot 31,\ 5^3) = 1,906,500$$

Of these, we see that the second set already has the correct lcm. The product of the numbers in this set is $1,953,000$.

Exercise 30

In base 2, a number is divisible by 3 if:

(**A**) its digit sum is a multiple of 3.

(**B**) its last two digits are 11.

(**C**) its alternating digit sum is a multiple of 3.

(**D**) the number of its digits is divisible by 3.

Solution 30

Because a formal proof is a bit cumbersome, let us use a random example such as the binary number 1101. Write it in expanded form:

$$1101 = 1 + 2^2 + 2^3$$

and use the fact that $2 = 3 - 1$:

$$1101 = 1 + (3-1)^2 + (3-1)^3$$
$$1101 = 1 + 3^2 - 2 \cdot 3 + 1 + 3^3 - 3 \cdot 3^2 + 3 \cdot 3 - 1$$
$$1101 = 3 \cdot (3 - 2 + 3^2 - 3^2 + 3) + 1 + 1 - 1$$
$$1101 = 3 \cdot (3 - 2 + 3^2 - 3^2 + 3) + 1 - 0 + 1 - 1$$

where in the last equation we have introduced a -0 term to emphasize the alternate digit sum. The derivation shows that any binary number the sum of a multiple of 3 and the alternate digit sum of the number. As a result, if the alternate digit sum of the number is a multiple of 3 the number is also a multiple of 3.

The other choices can be invalidated easily using simple examples. The correct answer choice is (C).

Competitive Mathematics Series for Gifted Students

Practice Counting (ages 7 to 9)
Practice Logic and Observation (ages 7 to 9)
Practice Arithmetic (ages 7 to 9)
Practice Operations (ages 7 to 9)

Practice Word Problems (ages 9 to 11)
Practice Combinatorics (ages 9 to 11)
Practice Arithmetic(ages 9 to 11)
Practice Operations (ages 9 to 11)

Practice Word Problems (ages 11 to 13)
Practice Combinatorics (ages 11 to 13)
Practice Arithmetic and Number Theory (ages 11 to 13)
Practice Algebra and Operations (ages 11 to 13)
Practice Geometry (ages 11 to 13)

Practice Word Problems (ages 12 to 15)
Practice Algebra and Operations (ages 12 to 15)
Practice Geometry (ages 12 to 15)
Practice Number Theory (ages 12 to 15)
Practice Combinatorics and Probability (ages 12 to 15)

This is a series of practice books. With the exception of a few reminders, there are no theoretical explanations. For lessons, please see the resources indicated below:

Find a set of free lessons in competitive mathematics at www.mathinee.com. Addressing grades 5 through 11, the *Math Essentials* on www.mathinee.com present important concepts in a clear and concise manner and provide tips on their application. The site also hosts over 400 original problems with full solutions for various levels. Selectors enable the user to sort essentials and problems by test or contest targeted as well as by topic and by the earliest grade level they can be used for.

Online problem solving seminars are available at www.goodsofthemind.com. If you found this booklet useful, you will enjoy the live problem solving seminars.

For supplementary assessment material, look up our problem books in test format. The "Practice Tests in Math Kangaroo Style" are fun to use and have a well organized workflow.

Made in the USA
Middletown, DE
04 January 2016